Why Women Should Rule the World

Why Women Should Rule the World

Dee Dee Myers

HARPER
An Imprint of HarperCollins*Publishers*
www.harpercollins.com

HarperCollins books may be purchased for educational, business, or sales promotional use. For information, please write: Special Markets Department, HarperCollins Publishers, 10 East 53rd Street, New York, NY 10022.

FIRST EDITION

Designed by Laura Kaeppel

Library of Congress Cataloging-in-Publication Data

Myers, Dee Dee.
Why women should rule the world / Dee Dee Myers.—1st ed.
p. cm.
ISBN 978-0-06-114040-2
1. Myers, Dee Dee. 2. Press secretaries—United States—Biography. 3. United States—Politics and government—1993–2001. 4. Clinton, Bill, 1946—Friends and associates. 5. Women in politics—United States. 6. Women in politics. I. Title.

E840.8.M94A3 2008
973.929092—dc22
[B] 2007047962

08 09 10 11 12 ID/RRD 10 9 8 7 6 5 4 3 2 1

To my mother, Judith Burleigh Myers,
whose love, encouragement, and congenital optimism
taught me to see opportunities, rather than obstacles.

Contents

Introduction

Women should rule the world.

That was it, the answer to my frustration and growing political alienation. It seemed so simple, so obvious. Women!

If we were in charge, things might actually change. Instead of posturing, we'd have cooperation. Instead of gridlock, we'd have progress. Instead of a shouting match, we'd have a conversation. A very long conversation. But a conversation nonetheless. Everyone would just hold hands and sing "Kumbaya."

Or would they? What would it be like if women ruled the world, I began to wonder? Would anything really change? Would the world be a better place? My hunch was that more women in public life would, in fact, make things better. After all, more women already have.

It's easy (and perhaps a bit facile) to argue that men haven't done such a great job. The last century was the bloodiest in human history, and so far, this one has been a tale of war, terrorism, religious extremism, abject poverty, and disease. I'm not saying it's all men's fault. But let's just say, they've been in charge, and it doesn't seem we're much closer to finding answers to these profound and vexing problems.

On the other hand, if there are societies where women have truly ruled, they are few and far between. For virtually all of history, woman has played a supporting role to man's, well, leading man. A comprehensive review of encyclopedia entries published in the early 1900s included only 850 women, though it covered a span of nearly 2,000 years. And the queens, politicians, mothers, wives, mistresses, beauties, religious figures, and women of "tragic fate" were notable mostly for their relationships with men.

I have always believed that women *could* rule the world. As far back as I can remember, it has seemed obvious to me that women were, in fact, every bit as qualified as men in most endeavors, and better than them at many. Of course, the corollary—that men are better than women at some things—also seemed obvious, at least after the sixth grade. Before that, I thought I could do anything any boy could do. I was a good student and a good athlete, and I didn't have much trouble keeping up with boys in the classroom or on the playground. But then Doug, another sixth grader at Wiley Canyon Elementary School in California, challenged my friend Peggy and me to a game of two-on-one basketball, first side to ten would win. He beat us 10-0.

I realized then that athletic boys are better basketball players than most girls, even the ones like Peggy and me who spent a fair amount of time shooting hoops. While I confess this was a bit disappointing at the time, I certainly didn't think that boys were better at everything, or even most things. That idea simply never occurred to me.

Maybe it's because I grew up surrounded by strong women. My mother, a product of her generation, left college after two years to marry my father, a young Navy pilot. Within a few years, she had three little girls and a husband who was often at sea. With Castro's ascent in Cuba, then the war in Vietnam, my dad was gone for weeks or even months at a time, and my mom was left to manage alone. One of my earliest memories is of helping my mom pack a little plastic Christmas tree, some cookies, and a few wrapped packages into a big box to send my dad, who was on a ship somewhere in Southeast Asia. But she never complained (at least not when my sisters and I were listening), and she never seemed overwhelmed by all that she had to do. The Navy, like all branches of the military, would collapse without the community of able women (and now a lot of men) who manage things stateside, while their husbands (and now some wives) are away. My mother and her network of Navy wives helped each other tend to sick children, unstop kitchen sinks, and deal with worrisome news from the war raging half a world away.

After my father left the Navy, we moved to the suburbs of Los Angeles, and my mom eventually earned both undergraduate and graduate degrees, then went to work, first as a counselor at a local college, then as an executive at the phone company. She was good at what she did, rose quickly in her various jobs, and got a lot of satisfaction from her professional accomplishments. I didn't always like it when my mom was gone, but I never doubted that what she was doing was important. At the time, most of the mothers in my neighborhood stayed home, so what my mom was doing was unusual. But my dad was supportive, and my sisters and I were more proud than displaced—even when we had to eat dry macaroni and overcooked hot dogs every time it was my sister Betsy's turn to make dinner. It never occurred to me that I wouldn't go to college and have a career—as well as a family—of my own. Both my parents, but especially my mother, encouraged me and led me to believe that it was possible.

My father's mother, Grandma Bernadette, also shaped my ideas about what women could accomplish, in ways I think she never would have imagined. Her husband—my grandfather—died of congestive heart failure (he'd had rheumatic fever as a child) when he was just thirty-seven, leaving her with five children: my dad, who was eleven, and his four sisters, ages twelve to two. My grandfather had owned a gas station on Main Street in Racine, Wisconsin, while my grandmother was busy raising the children and playing the organ at St. Joseph's Catholic Church. She hadn't been very involved in the business—and it certainly wasn't a business where one expected to find women in 1946. Because of his heart condition, my grandfather didn't have any life insurance, but his business was insured. So when he died—as my grandmother liked to tell it—the insurance men came to her house, suggested she sell it and the gas station, and move with her children into the Catholic orphanage across town. She told them to get the hell off her porch and never come back. She kept the station and managed the day-to-day operations until she sold it more than thirty years later. She raised five children, put them all through college, and still found time to play the organ at Mass every weekday and five times on Sunday. While she clearly missed things about being married—and having a father for her children—she never really dated or considered marrying again. She would sometimes say she never found the right fellow, but her daughters believe that she simply liked being the boss.

So my grandmother—by fate, rather than design—was a small-business owner and single mom long before women routinely did either, let alone both. And I've often wondered: What would have happened to another family if the mother had died and left the father with five young children? How many men could have managed to run the business, raise the kids, and volunteer at church six days a week, all by themselves?

In addition to my mother and grandmother, I grew up surrounded by accomplished women. The principal of my elementary school. My guidance counselor in high school. My father's sisters. My friends' mothers, and my mother's friends. It seemed to me that women were capable of doing just about anything. Not that they were always allowed to, of course. When I was in second grade (even before I learned that boys were better at basketball), our teacher asked us to draw a picture of what we wanted to be when we grew up. The kid next to me, Robert, drew himself as a TV repairman. While his choice of career may not have thrilled his parents, it struck me hard. *Wow*, I thought. *He can be anything. I have to be a teacher, or a nurse, or a nun.* I drew myself as a teacher.

Happily, the years since I finished the second grade have seen an exponential increase in options. Girls can now aspire to be elementary school teachers or university presidents; nurses or doctors; nuns or—in many denominations—priests or ministers or rabbis. Girls *and* boys can be engineers, entrepreneurs, or astronauts. They can repair televisions or appear on them as actors or journalists. They can build homes or stay home with the kids.

And they can be press secretary to the president of the United States, as I was.

When I first started working in politics, as a junior aide on Walter Mondale's 1984 presidential campaign, it never occurred to me that I would one day work in the White House. There were plenty of women among the volunteers who stuffed envelopes and walked precincts. But there were fewer and fewer on each successive level of influence and access. In the subsequent years, the numbers increased, as I moved from job to job—in the California state legislature, for the mayor of Los Angeles, on the gubernatorial campaigns of Tom Bradley and Dianne Feinstein, and on the presidential campaign of Michael Dukakis. But electoral politics was still very much a white boys' club.

When I joined Bill Clinton's start-up presidential campaign in 1991, I was confident that women would play an ever more important role, but I never gave a minute's thought to what would happen if we won. When we did—and I became the first woman to serve as White House press secretary—it changed my life. But it didn't change the world. And I came to believe that it would take more women—lots more women—to do that.

After I left the White House, I kept a foothold in the business of American politics: as a talk-show host, analyst, commentator, speechmaker, and occasional writer. I was no longer a practitioner, but I was still a partisan, a Democrat, a blue-stater through and through. And I enjoyed the give-and-take of the political debate. But over the years, something changed, and I found myself more and more frustrated by the bitterness that now gripped the capital. Increasingly, it seemed, both sides were more interested in winning the argument than solving the problem. And the result was gridlock, polarization, and cynicism.

Surely there was another way, a better way. And I started to think about how we might move from a culture of confrontation to one of consensus, from I-win-you-lose to win-win. Was anyone in Washington practicing what I was only preaching? Were there people talking *and* listening to each other? Were they working together? Were they treating each other with respect and trying to see the world through each other's eyes? And I realized that, yes, there were some. And one of the places it seemed to be happening on a regular basis was among the women in the U.S. Senate.

Now, granted it's still a relatively small group: sixteen women. And it's easier to find comity among sixteen than among 100 or 535 or 300 million. But something seemed to be happening there. On paper, the women didn't have that much in common. They were liberal and conservative. They came from small states and big ones, both coasts and the middle. Several were single; others were mothers and grandmothers. They had different interests, different agendas, and different strengths. And yet. They had managed to transcend the bitter partisanship that has infected much of Congress, and forged not just political alliances on issues where they agreed—but genuine friendships.

"We relate on a personal level, because every one of us has had to overcome the obstacles of people underestimating us and people trivializing us," said Senator Kay Bailey Hutchison, Republican of Texas. "We're good friends."

The ideologically diverse group has never formed an official caucus, but in recent years, they've worked together on a variety of issues, including more access to individual retirement accounts for homemakers, more funds for home health care and breast cancer research, and a resolution condemning the ruling military junta in Burma for its brutal suppression of pro-democracy demonstrators. They have also reached out to other women leaders around the world. A few years ago, they met with women leaders from Northern Ireland, who were working to build a more civil society in that war-torn country; the Irish women came away inspired.

"My experience has been that women tend to be better at

working across the aisles and are more pragmatic and results oriented," said Senator Susan Collins, Republican of Maine.

While sixteen women in the Senate does not an airtight argument make, it certainly reinforced my own prejudices. Women do seem more interested in consensus. They do seem less consumed by the constant who's-up-and-who's-down score-keeping aspect of the political game. They do seem more willing to listen to other people's opinions. That's not to say that all women fit this model; they don't. But wouldn't increasing the number of women in Congress change the culture? Wouldn't it make the elusive search for common ground more fruitful? Wouldn't it make the political process more productive?

Wouldn't it? Yes, I thought; it would. In fact, if there were more women in positions of power, not just in Congress, but across the United States and around the world, lots of things would be better. Not perfect. But better. We'd have more representative government; a stronger economy; and a healthier and more sustainable planet. We'd be better able to resolve conflicts and keep the peace. We'd have stronger families.

And so I set out to write this book: *Why Women Should Rule the World.*

I knew my own story, as political operative—and as a mother, a daughter, a sister, a wife, and a friend. But I needed more. So I talked to friends and read articles, studies, and books. I interviewed prominent and successful women, from primatologist Jane Goodall and Senator Dianne Feinstein, to activist and skin care entrepreneur Anita Roddick and Nobel Prize laureate environmental activist Wangari Maathai. I explored the growing body of scientific literature on the topic.

When I actually sat down to explore the argument, however, I realized it was going to be harder than I thought. Women haven't been able to carve out much space on the top floors of any endeavor, in any country or culture in the history of the world. Without a doubt, they've made tremendous progress in the past three decades, but the numbers are still small. In the United States, millions more women than men vote, and we have a female Speaker of the House for the first time in history. Still, women make up only 16 percent of the U.S. Senate, 16 percent of the House, and not quite 24 percent of state legislators. Only eight of the nation's fifty governors are women. And while a woman has finally made a serious run, no woman has ever been elected president. Around the world, there is an increasing—if still small—number of women serving as heads of state or heads of government; but the small numbers make it hard to predict just how things would change if in every region of the world, every level of government was half women.

Ditto business. Women make the vast majority of consumer decisions in this country—by many accounts, more than 80 percent. But we still don't have enough influence at the top of the corporations that make and sell those goods and services. True, women now fill about half of all managerial positions, but among Fortune 500 companies, women account for only 16 percent of corporate officers, 5 percent of top earners—and an anemic 2 percent of CEOs. Is it really possible to know how the world would change if women had their names on half the doors to the executive suites?

The pattern repeats and repeats. Women make up half of law school graduates and roughly a third of all lawyers. But they account for only 15 percent of partners in law firms or federal judges, and 10 percent of law school deans or general counsels at Fortune 500 companies. Women make up nearly half of medical school graduates—but only a quarter of doctors and 10 percent of the deans of medical schools. They are 20 percent of university presidents, but still woefully underrepresented in tenure-track teaching positions, especially in math, science, and engineering. How would a giant increase in the number of women at all levels change law, medicine, and academia?

These were among the questions that I wanted to explore.

Of course, the questions run deeper than the statistics that quantify women's achievements—or the lack thereof. In the past couple of decades, there has been a mountain of research and commentary on the relationship between gender and just about everything—from leadership style, to ethics, to sex drive. And as the volume of information has grown, so too has the volume of the debate about what it means. Are the alleged differences real? Which, if any, are innate? Which are the result of socialization? And how do they affect expectations about gender roles?

As I began looking into questions like these, I was struck by the ferocity of the debate that still surrounds the "nature versus nurture" question. On the nature side, an eclectic group of scientists, philosophers, polemicists, parents, and religious traditionalists believe that sex roles are genetically, even divinely, determined. According to this view, women are nurturers, designed to have and raise the babies, while men are programmed to compete in the world and support their families.

The arguments are equally passionate on the other side, where a committed assortment of psychologists, sociologists, feminists, parents, and progressives argue that nurture is the root cause of behavioral differences between men and women. While the stack of studies is getting bigger, they claim that the evidence linking biology and behavior is tenuous; culture, they say, is the culprit. From infancy, boys are expected and encouraged to behave one way, girls another. And to see it any other way is to open the door to the kind of biology-as-destiny limitations that have held women back for thousands of years. If women are more nurturing, if they're better at relationships, isn't it also possible that men are better at, say, math or science? And wouldn't that then justify the dearth of opportunities for women in those fields?

These, too, became questions I wanted to explore.

I'm not a sociologist. I'm not a psychologist, or a biologist, or a political theoretician. But as I began this book, I wanted to try to paint a picture, in laymen's—or should I say laywomen's?—terms, of what changes when there are more women in positions of power and authority across public life. And I hoped—and let's be honest, expected—the results would make it obvious that the influence of women has been an overwhelmingly positive thing. Not because women are the same as men, but because of the many ways they are different.

At the same time, I realized that an honest look at the upsides of empowering women would also require me to look at the obstacles, from the big cultural, historical, and biological forces, to the challenge of balancing work and family and the internal barriers that keep women from being all that they can be.

This book is not an attack on men. It's not meant to demean or marginalize them. After all, my father is a man. I'm married to a man. I gave birth to a baby man. I think men have done wonderful things, from inventing the wheel (though it may have been a woman's idea, but somehow a man got credit), to walking on the moon. Truly, the list of man's (and I don't mean "mankind's") accomplishments is so long and so profound that it seems silly to try to quantify it. But that doesn't mean the world wouldn't be better if there were more women in public life. If women had more power, not just in the United States, but around the globe. If women had the same access to education and economic resources and health care. If women had equal rights and equal opportunities. If there were more women in boardrooms, and classrooms, and operating rooms and courtrooms. If women's ideas and opinions and life experiences were accorded the same weight as men's. If girls were as valued as boys.

If women ruled the world.

Part I

Why Women Don't Rule the World

Chapter 1

Between a Political Rock and a Promise

"When a man gets up to speak, people listen, then look. When a woman gets up to speak, people look; then if they like what they see, they listen."

—Pauline Frederick

Just six days before Bill Clinton was sworn in as the forty-third president of the United States, he announced that I would become White House press secretary—the first woman, and at thirty-one, one of the youngest people ever to hold the job. Oddly, I can hardly remember the exact moment when it happened. I remember being at the Old State House along the banks of the Arkansas River in downtown Little Rock. I remember being surrounded by many of the friends and colleagues I'd worked with on the campaign across the previous fifteen months. And after checking old newspaper clips, I know that the president-elect introduced a handful of top staff himself, before turning the podium over to Mack McLarty, his newly appointed chief of staff, to fill out the roster of once and future aides, including me.

But like so many of my memories from that time, this one feels a little freighted. A moment that might have been grand and indelible and joyful is wrapped in strands of more complex emotion, like uncertainty, trepidation, and disappointment.

Of course, I was thrilled at the prospect of working at the White House. It was something I had dreamed about since I caught the political bug in college. A few years ago, my friend Red produced a letter I had written to him from France during my junior year abroad; the return address was "1600 Pennsylvania Avenue." But I never really believed it would happen—certainly not as the White House press secretary. Yet here I was, getting ready to accompany a young and energetic new president to Washington. As part of the original campaign team, I felt I had a stake in this extraordinary journey, and I was eager to take the next leg.

But there was so much more to this particular story.

When I went to work for Bill Clinton in December of 1991, he was the longest of long shots. George H. W. Bush was still popular in the wake of the Gulf War, and it seemed impossible that a cocky, young governor from a poor Southern state could win the Democratic primary, let alone defeat an incumbent. But I didn't care. I'd met Clinton a few times over the years, and he impressed me. He was talking about things that I thought Democrats needed to talk about, from heath care and welfare reform to lifelong education for workers to keep them competitive in a changing economy. When I interviewed with Clinton for a job as the campaign's national press secretary early that fall, he didn't ask me a single question about my background or qualifications; we focused, instead, on his vision for the country. Bill Clinton knew why he wanted to be president. Fifteen minutes into our conversation on a car ride from the Hollywood Hills to the San Fernando Valley, I was sure I wanted to work for him. A few days later, I was offered the job.

As I joined the campaign, I thought I had little to lose. Four years earlier, I'd been a state press secretary in the Dukakis campaign; now I would be national press secretary. It was a huge step up in responsibility and visibility, and I knew I'd gain valuable experience. If Clinton lost, I'd be well positioned for the next presidential cycle four years later, when there would be no incumbent. And if he won? Well, my mind could hardly go there.

I spent most of the next year on an airplane, as Clinton survived a series of political near-death experiences, won the Democratic nomination, and was elected president of the United States.

Winning was a new experience for me, as virtually every candidate I'd ever worked for had lost. I remember standing in a bar somewhere in Little Rock on election night—basking in the novel glow of victory, cocktail in hand—when it occurred to me: We have to work tomorrow! So I rounded up the members of my team, and we came up with a plan to staff the press office beginning at eight o'clock the following morning. Our department alone got nearly 900 phone calls in the next two days.

The long and competitive campaign had taken a toll on the candidate, the staff, and the reporters who chronicled our every move. But as soon as the race ended, we had to dig out from under our exhaustion and start a new, equally intense mission: preparing Bill Clinton to assume the presidency at precisely twelve noon on January 20, 1993. There had been some transition planning at the top levels of the campaign in the weeks leading up to the election; after all, Clinton had led in the polls since the Democratic convention in July. But it had not trickled down to many of the people who would be expected to run the day-to-day operations of a president-in-waiting.

It took more than a week for the transition staff to be officially named. The delay was discomfiting, but the actual news did little to clarify my future. I was told that I would be press secretary to the transition, but that I shouldn't infer from that that I would be White House press secretary. The blunt message stunned me. I certainly didn't expect that my role in the transition would guarantee me a particular job in Washington. And while I wanted to be press secretary—and I'd earned my stripes over the previous decade, particularly during the recently ended campaign—I knew I didn't have the perfect résumé. I was thirty-one years old, I had never lived or worked in Washington, and I was a woman.

As November became December, I tried not to focus on the personnel rumors that swirled in the absence of any real news. *Myers is in. Myers is out. Clinton doesn't know what to do about Myers.* Bit by bit, Clinton pieced together his cabinet, unveiling his choices in a series of press conferences. But as the holidays approached, the White House staff remained a missing part of the puzzle.

At some point, George Stephanopoulos, who was heading up communications planning, came to talk with me about jobs. When he asked me what I wanted to do, I told him I was hoping to be named White House press secretary. But I was shooting high, laying down an opening bid as much as expressing a real expectation. In truth, I thought I could just as easily end up as either the chief spokesman at an agency or department, or as a deputy press secretary in the White House. In either case, I hoped that after a year or two, I would get a shot at the top job. I knew Warren Christopher a bit. When he was named secretary of state in late December, I toyed with the idea of working for him. Like me, he was from Los Angeles, and we flew home together on Christmas Eve. But unlike me, he was escorted off the plane, onto the tarmac, and into a waiting limousine by State Department security agents when we landed; I met my dad at baggage claim.

In early January, I met with Stephanopoulos and Ricki Seidman, another member of the transition team, in the basement of the Arkansas governor's mansion, where Bill and Hillary Clinton's personal staffs were working. Rather than the grand, or at least formal, invitation to join the White House staff that I had imagined, George and Ricki pulled me aside in a hallway and told me the plan: I would have the title of White House press secretary. But the job would be a little different. George would be director of communications; he would handle the daily briefings, as he had during the transition, and I would be the backup briefer. He would take the press secretary's office in the West Wing; I'd have a smaller office in the same suite. He'd carry the highest rank of assistant to the president (as all previous press secretaries had); I'd be a deputy assistant—a lower rank that came with a smaller salary (natch). I didn't know what to say.

I'd just been told I was going to be White House press secretary—a job that I had barely dared to dream about—but it wasn't quite what I'd imagined. Suddenly, I found myself staring down the barrel of a predicament that I knew was all too common among women: responsibility without corresponding authority. Most memorably, I'd seen this happen to Susan Estrich during the Dukakis campaign four years earlier. She'd been named campaign manager in the aftermath of a silly scandal that had nonetheless cost both the campaign manager and chief consultant their jobs. Susan was brought in to repair the damage and run the operation. And despite her talent and best efforts, the powers-that-be never really gave her the reins. Nonetheless, she got saddled with a load of blame for the campaign's failures—even the ones that had been beyond her control. As I watched her, I vowed I would never let that happen to me. Yet here I was. And I was pretty sure I knew why.

THE "BEAN COUNTERS"

Bill Clinton liked me. I'd come to the campaign early, and for more than a year, I'd been at his side nearly every day. He knew my strengths, my weaknesses—and the limitations of my experience. He wanted to find a place for me in the White House, even if he wasn't sure I was quite ready to step into the press secretary's job. Still, he had a political problem to solve.

Throughout the campaign, Candidate Clinton said he would appoint record numbers of women and minorities to high posts in his new administration; he wanted a government that "looked like America." And he implied that one of the "Big Four" cabinet jobs—state, treasury, defense, and justice—would go to a woman. But after he announced his pick for the first three of the Big Four—and none was a woman—Clinton found himself playing defense in the quota game that he'd helped to start. As the pressure mounted, Clinton lashed out at the "bean counters"—the feminists and their liberal allies—who were tracking his every move, determined to hold him to his promises. As the search for a woman to serve as attorney general foundered, Clinton could ill afford to be seen as throwing over a loyal woman (me) in favor of a man. He was caught between a political rock and a promise.

I agreed with Clinton's goal: a government that looks like America is important, and I believed then (and believe now) that the country is well served when people in positions of power work a little harder to recruit from the enormous pool of talented women. Still, I didn't want to be a temporary solution to a longer-term problem. I didn't want to be set up to fail so that Bill Clinton could kick his personnel problem down the road.

"I can't do it," I told George. "This is a prescription for failure."

"You have to," he said. "The president of the United States is asking you to serve. You can't say no." Ricki agreed.

"It'll never work," I argued.

But I wanted it to work. I wanted to serve the new president. I wanted to be White House press secretary. So in spite of my reservations, I accepted the job as offered.

As I left the mansion, I remember feeling anxious, even a little embarrassed, by my situation. On the one hand, I was going to work in the White House! I'd only been in the building once, when President-elect Clinton paid a visit to President Bush shortly after the election. In the wake of a bitter campaign, we were about as welcome in the Bush White House as skunks at a garden party, but I was still awed by the place. It was less grand than I had imagined; the rooms were a little smaller, the carpets a little more frayed. But everywhere you looked, the backdrops were as familiar as the events that had happened there: the white column–lined colonnade in the Rose Garden where I'd seen presidents walk for as long as I could remember. The podium in the briefing room, backed by the familiar blue drape and White House plaque. The Norman Rockwell paintings hanging on the wall along the ramp that leads from the press work space to the West Wing offices. I tried to memorize every detail—even as the Bush White House aides herded me unceremoniously away from the cluster of reporters after a brief appearance by the president and the president-elect outside the Oval Office. "No spinning in the Rose Garden," they kept shouting, as though this were an iron-clad and well-known rule. I was shown to the West Lobby, where one of the ushers gave me my first little white box of M&M's emblazoned with the presidential seal. (I still have it, though I'm sure I ate dozens, if not hundreds, of others in the years that followed.)

As exciting as the new digs were, the circumstances were less so. I knew that in order to protect any chance I had at being effective, I had to try to ignore the obvious downsides of the arrangement; I had to try to make it work. But I also knew the most skeptical audience would be my primary constituency: the press. Reporters—even those with whom I had friendly relationships—were going to care more about protecting their own interests than in cutting me any slack. Since the arrangement would raise questions about how much access they were going to get, I knew they weren't going to like it.

Before my appointment was made public—and just days before we were to leave Little Rock for pre-Inaugural festivities in Washington—I was told how many people I could hire (fourteen) and how much I could pay them (not much). Given that the press office was expected to be ready to respond to every need of the White House press corps the minute—literally—that the new president swore the oath of office, I had to work fast. All but one of my staff came from the campaign team that I had pulled together in bits and pieces over the previous year. And like me, most of them also lacked Washington experience. But they were loyal to Clinton—and to me—and I knew I was going to need friends.

A few days later, I joined my colleagues at the Old State House for the public announcement, determined to make the best of a difficult situation.

But I was right to be worried. In the first months of the new administration, questions about the press office swirled, with my decreased stature as a central concern. Several months later, Mark Gearan replaced George as communications director, and my role expanded to include the daily briefing. It was an important step forward, but it still didn't resolve questions about my access. And those questions would persist throughout most of my tenure.

At times, I felt I had the support of the president and the senior staff, as well as access to the information I needed, to do my job. And at times, I didn't. I didn't have as much experience as some of the men who came before me. And too often, that became the justification for limiting my role, which in turn guaranteed that I'd be less effective. The circular logic (and its very real effect) was infuriating, and at times I struggled to control my anger. I'm still not sure what I could have done to change things.

Mine is a cautionary tale. The seeds of so many of the obstacles I faced were sown during the decision-making process that ended with my being offered the job. And more than anything else, that process shaped my years as White House press secretary. I had the same responsibilities—but less authority. And fewer of the trappings of power—the office, the rank, the money—routinely accorded to previous press secretaries. Early on, I tried to argue that those trappings said more about Washington's obsession with symbols than they did about me, but I wasn't very convincing. The president and the senior staff made the job less important than it had been. And that made *me* less important.

And so goes the downward spiral. Sometimes, just putting a woman in a job can make it seem less important. In fact, women devalue whole sectors of the economy just by showing up; studies show that both men and women attach less prestige to certain professions if they have more women—or are expected to have more women in the future. For instance, doctors used to be among the most trusted and respected of professionals. They were well thought of and well compensated, often living in the fanciest neighborhoods and driving the fanciest cars. But that's starting to change. Surely the economics of medicine have changed. But so, too, has the prevalence of women. In 2002, 40 percent of medical residents were women—a fact that supports predictions that by 2010, 40 percent of all U.S. physicians will be women. And while doctors are still respected, these forces, both economic and social, have started to chip away at their earning power—and their prestige.

Not only can the presence of women devalue certain jobs, but often it's not until the job is devalued that women are even hired. Take Katie Couric. Only when the network evening newscasts had become floundering franchises—with both viewership and influence flagging—did they let a solo woman into the anchor club. (Barbara Walters and Connie Chung had been allowed to serve as co-anchors in previous decades, but both job-sharing experiments had failed.) Now, as the ratings of the *CBS Evening News*—a program that already seems conspicuously out of date—continue to sag, Couric is on the receiving end of some harsh criticism. True, she was paid tens of millions of dollars to turn things around, and the network is understandably disappointed. But her male counterparts at ABC and NBC are also highly paid—and their broadcasts haven't shown any real ratings growth, either.

It's a no-win situation. For too many women, the job becomes less important because they're in it. They're given fewer resources—like stature or staff—to achieve the same results. And when they don't hit the mark, they get blamed. Before I went to work at the White House, I was aware of this dynamic; I'd watched it happen to other women. But when it happened to me, the punch it packed still surprised me. It took me years to realize that it wasn't all my fault.

"He's Got a Family!"

About a year and a half into my tenure, I found out—from a news story about staff salaries? I don't quite remember—that one of the deputies in another office was making more money than I was. Like me, he was a deputy assistant to the president, a title that carried a maximum salary of $110,000. He was making the max; I was making $10,000 less. But I had more responsibility than he did; by any possible measure, I outranked him. And yet our paychecks didn't reflect that. I decided I deserved a raise, and I went to see Leon Panetta, who had recently become chief of staff. I explained the situation and said that I wanted a $10,000 raise. I had one of the more demanding jobs in the White House, I argued; I should be at the top of the pay scale in my classification.

"He was a partner in a law firm," Panetta told me, coolly. "He took a huge pay cut to come here."

"I appreciate that," I said. "I'm sure he'll more than make up for it when he leaves. But that's beside the point. I'm not asking you to cut his pay. I'm asking for a raise."

"No," Panetta said flatly. "I don't have the money."

"But it's not fair," I countered. "I have more responsibility. My pay should reflect that." *This isn't a heavy lift*, I thought; *I'm only asking for about $800 a month.* Eight hundred dollars! But there wasn't a smidgeon of give in Panetta's position.

"Look, we have to pay people based on previous experience and salary history. Plus, he's got a family. It's not going to happen." The meeting was over. I couldn't believe what I'd just heard. And I was livid. As I walked back to my office, I reviewed my options. I could fight it, I thought. But who would I appeal to? The president? I knew I would never waste his time on a $10,000 salary dispute. I could take formal action. I wasn't sure what that would entail, but I assumed it would take time, something I didn't have. I also knew that if it leaked, it could become a political problem for the president. Again, that was something I was unwilling to do. Under the circumstances, I thought my only real option was to let it go, to just accept the injustice and get back to work.

Before the meeting, I thought my case was pretty straightforward, and I thought the best way to deal with it was simply to go straight to the boss and lay out the facts. I wasn't sure I'd get the raise, but I assumed that I'd get a sympathetic hearing; at a minimum, I thought Panetta would concede I had a point and promise to look into it. After all, he had quit the Nixon administration and switched political parties after a dispute over civil rights enforcement. After the meeting, I realized I should have been prepared for a fight. I could have surreptitiously surveyed other women on the staff to see if there was a pattern of discrimination. Or recruited some allies to help argue my case. Or found out what appeals might be available to me if he said no. It just never occurred to me that I would have had to. What's more, I had learned by then that you have to pick your battles pretty carefully; you have to decide "which hill you want to die on," as my mother would say. And at the end of the day, this wasn't the hill I wanted to die on. That would come later.

Still, it's a hill too many women get stuck on. More than four decades after President Kennedy signed the Equal Pay Act into law, studies show that women are still paid less than men for doing the very same jobs. A comprehensive study by the U.S. government's General Accounting Office (now the Government Accountability Office) found that in 2000, women working full-time earned just eighty cents for every dollar men earned. The study took into account a wide range of factors, such as education and work experience, that might affect earnings; without controlling for those variables, the data showed that women earned 44 percent less than men.

Other studies report similar findings. Hilary Lips, a psychology professor at Radford University in Virginia, has found that even when you control for other factors, women still make less than men in virtually every job category and every field. Only in jobs with salaries from $25,000 to $30,000 were men and women paid the same. And the more that certain jobs paid—or the more education they required—the more the gap actually grew. For example, Lips found that women psychologists earned eighty-three cents for every dollar their male colleagues earned, women college professors seventy-five cents, and women lawyers and judges just sixty-nine cents. Women elementary school teachers earned ninety-five cents for every dollar earned by men, women bookkeepers ninety-four cents, and women secretaries eighty-four cents. "It cannot be explained in any way except that people think that what men do is more important and more valuable than what women do," Lips concluded.

Need more evidence? Recently, when I signed onto AOL, I found a headline that read: "Best Jobs by Gender . . . Women

Make the Most as Nurses." When I clicked on the item, I found that nursing was indeed the highest-paying female-dominated profession, with an average salary of $56,900. But the average male nurse makes $64,200! In fact, in all five of the highest paying female-dominated jobs—including human resource manager, executive assistant, paralegal, and accountant, in addition to nurse—men earned more.

And it's not that men are necessarily more ambitious. In 2005, Catalyst, a research and advisory organization that tracks women in business, talked to 950 top executives, both men and women. Fifty-five percent of women, including women with children, said they wanted to be CEOs—virtually the same as the number of men who said they wanted to run the company.

If that's true, why do so few women make it to the corner office? According to Catalyst, a quarter of the women said they lacked operational experience—the kind of profit-and-loss responsibility that is a key to success in the business world. But nearly twice as many said that it was because they were shut out of the informal networks—the golf games, poker games, and men's clubs—where information gets exchanged, relationships get established, and careers get launched. And so it is that women are judged by their performance—while men are judged by their potential.

Over a lifetime, little differences become big differences—and the cumulative effect can be eye-popping. A few thousand dollars in an entry level job can add up to hundreds of thousands of dollars over the course of a career. An executive at Boeing told *Fortune* magazine that in every job category at the company, men made more than women. And when the raises were doled out, things only got worse. He explained that even if you subtracted 2 percent from the across-the-board 7 percent raise for a man making $100,000 and added 2 percent to a woman making

$50,000, his raise would be $5,000, hers $4,500—and the gap would widen. "There was never enough money to fix the problem," he explained. "It was heart-wrenching to figure out how to bring the women up and at the same time not penalize the men. Eventually, we just gave up."

Surely, one reason women make less is that too often, they don't ask for more. (And I suspect—based on my own experience—that even when women ask, their bosses find it easier to tell them no.) Research shows that women are likely to accept salary offers, no questions asked. One study that tracked graduate students from Carnegie Mellon University found that the men negotiated their initial salary a staggering eight times more often than the women. As a result, their starting salaries were on average $4,000 higher—a 7.4 percent difference that would surely grow and grow and grow, like compound interest, over time.

Linda Babcock and Sara Laschever, authors of *Women Don't Ask: Negotiation and the Gender Divide*, write that by failing to negotiate that first salary, a woman (or less often, a man) stands to lose more than $500,000 by the time she hits sixty. So why don't women pipe up? They're socialized not to, says Babcock. "Society really teaches young girls, from the day they are born, to think about the needs of others and not to think about their own needs. So they grow up not thinking about themselves or how to get what they want, but only thinking about others, really. In addition, women are also penalized for being too aggressive. When a man drives a hard bargain, he knows what he's worth and by God, he's going to get it. But when a woman does the same? She's a pushy broad, and no one wants to work with her."

UNINTENDED CONSEQUENCES

At the White House, devaluing the press secretary's job created its own set of problems. Eventually, there were four of us with overlapping responsibilities: David Gergen, counselor to President Clinton, who had served three previous Republican presidents in a variety of communications-related functions; George Stephanopoulos, who had become senior adviser to the president for policy and strategy but continued to talk regularly with reporters after he changed jobs; Mark Gearan, who succeeded George as communications director; and me. It didn't take a rocket scientist to recognize that this was a bad setup, especially for a press secretary who was the only woman and the youngest member of the group.

But it was also bad for the president. It was my job to stand behind the lectern in the briefing room every day and speak on behalf of the president, the White House, the country. I needed to know not just the details of policy, but also the state of play. I had to be able to steer reporters away from bad information—and to shape realistic expectations about what might happen next. But too often I had trouble getting the information I needed to do all that; too often, I wasn't in the room. It wasn't that people intentionally cut me out. To a great degree, I had good relationships with my colleagues, and they trusted that I was a loyal member of the team. But if one of the other three—Gergen, Gearan, or Stephanopoulos—knew something, people often assumed that the "press office" had been informed and that I knew it, too. In the crush of daily events, the handoff didn't always happen. And even when it did, a secondhand report could rarely take the place of an eyewitness account. So there were gaps in what I knew.

And sometimes those gaps became a source of missed opportunities, accidental misinformation—or damaging mistakes.

The worst such mistake involved a failed plot to assassinate former president George H. W. Bush while he was on a trip to Kuwait early in 1993. When evidence suggested that Saddam Hussein was involved, President Clinton ordered the FBI and CIA to investigate.

Over the next couple of months, I received regular updates about the investigation from the National Security Council staff, and once or twice a week, reporters would ask me about it at my briefing. My response was generally, "The FBI is still investigating. When they've finished their work, they'll forward their conclusions to the president, and he'll decide on a course of action." On Friday, June 25, an NSC staffer told me that the report was in its "final phases" and that the president would be receiving it soon. As it happened, I was asked about it at my briefing that afternoon.

"We hear that it's in its final phases, but it's not complete," I told the assembled reporters. A few minutes later, discussion again turned to the assassination attempt.

"Did you say that the FBI report on the Bush assassination thing is expected in the next few days?" I was asked.

"No, I said I understand it's wrapping up, but we don't have it yet."

"Do you expect it within days?"

"I don't know," I answered.

It turns out there was a lot I didn't know. The president had already received the report, and it established a clear link between Saddam Hussein and the plot. After consulting with his national security advisors, including General Colin Powell, who was then chairman of the Joint Chiefs, he had decided to launch a cruise missile strike against the headquarters of the Iraqi intelligence service in downtown Baghdad. To minimize civilian casualties, the strike was scheduled for the predawn hours on Sunday morning—early Saturday evening, DC time. But when I left the office on Friday, I didn't have a clue.

As usual, I got to the White House around nine o'clock on Saturday morning. I helped prepare the president for his weekly radio address at ten, and then spent a couple hours in my office catching up on paperwork. It was a quiet day; the president's only plans were to go for a jog and then play golf. As I headed out a little after noon to get my hair cut, I ran into David Gergen. "You'll want to be back around three," he told me. "And you'll want to look nice." *What in the world did that mean?* I wondered. I was dressed casually, as was the weekend custom in the Clinton White House, and there was no public schedule that afternoon, no news in the offing. When I asked him what was up, he said with just a hint of a smile, "I can't tell you." Given his demeanor—and the almost accidental nature of my invitation—I certainly didn't think it was a matter of international import. I got my hair cut, changed into business clothes, and headed back to the White House around three.

When I got there, Gergen brought me up to speed. In about an hour, Tomahawk missiles would be launched from two ships, one in the Red Sea, one in the Persian Gulf. Between launch and landing, President Clinton would call congressional and world leaders—including former president Bush. And once we received confirmation that the missiles had landed, the president would address the nation from the Oval Office. I was immediately torn in two directions. On the one hand, we had a ton of work to do over the course of the next several hours: Notify the press—and by extension, the world—of the president's pending address. Organize briefings by senior officials to provide further explanation. Compile a detailed timeline of events leading up to the strike. On the other hand, I was building up a righteous head of anger. I knew I'd misled reporters—unintentionally—the day before, and at some point, I was going to have to answer for it. Surely, this could have been prevented. George Stephanopoulos had known. David Gergen had known. But once again, I hadn't known. By keeping me in the dark, my colleagues had meant to protect me—and the mission. If a reporter asked me a tough question—and I knew what was coming—I might be forced to lie or to compromise the effort by revealing too much, the thinking went. But no one ever thought to ask what I'd already said.

A while later, my assistant, David Leavy, came to see me. The lower press office had announced a "lid," meaning there would be no more news or announcements from the White House that day, and the reporters who hung around the briefing room "just in case" had scattered in their usual Saturday afternoon diaspora. I went to see Mark Gearan; we decided we couldn't call the press back yet without tipping them off that something was happening. My anxiety deepened; I was going to have to answer for this, too. But right now, the only thing I could do was put my head down and do my job. The first round of stories would focus on the strike itself; questions about process could wait until tomorrow.

By virtually every measure, the assault was a success. Twenty-three of the twenty-four missiles had launched successfully. The president had received unqualified support from Russia, Western Europe, and even normally neutral Sweden. And the administration had sent a clear signal to would-be sponsors of terrorism: There will be consequences. What's more, news of the operation hadn't leaked—a huge victory in a White House that was already famously indiscreet. The general mood in the West Wing was euphoric. The new president and his team had handled their first real military action capably. There was little concern about a relatively minor misstatement from the podium. Unless, of course, you had done the misstating.

The following day, Doug Jehl from the *New York Times* called. He was writing a piece detailing the White House's success in keeping the missile strike secret. Only five White House aides had known the details in advance, and keeping the circle small had preserved the secret. But hadn't it also led me to provide erroneous information at my briefing, he wanted to know? Hadn't it also made the White House look less than candid?

I went to see National Security Advisor Tony Lake and Gergen. My misstatement wasn't good for the president, it wasn't good for the White House, and it certainly wasn't good for me, I argued. I needed to be able to tell the press that we were going to change some of our procedures to protect against similar mistakes in the future. They agreed. The only thing left was to talk to the president. On Monday morning, I caught him as he arrived in the Oval Office. I explained my situation. "They'd be on us if the story leaked, and they're on us 'cause it didn't leak," he said with more than a hint of irritation. But he agreed that we needed to make some changes.

The rest of the morning was taken up with a cabinet meeting, planning for our first meeting of the G7 a few weeks later—and preparing for a briefing that I thought might get personal.

Just before two o'clock, I headed to the podium prepared for the worst. After a few routine announcements, I opened the floor to questions. The first topic was the political impact of the weekend's events. When I shot down a question about whether politics had played a role in the president's decision to launch the missile strike, one of the reporters shot back, "How do you know that since you—by your own statements—were not part of any meetings having anything to do with this?"

The first punch landed hard. And it went downhill from there.

"Dee Dee, you've been reported to be concerned about the impact of the weekend on your credibility and on your office because of the erroneous lid that was put on and because of the statement you made Thursday [*sic*], not knowing, apparently, some of the things that were going on behind the scenes. Can you talk a little about your concerns?"

For the next thirty minutes, I answered question after uncomfortable question about my own role. I tried to explain that I'd made an "honest mistake," that we would "do things differently" in the future to prevent another such episode, that I had talked to senior officials, including the president, and that they had agreed. But there was blood in the water—all mine—and the sharks were circling. Finally, Helen Thomas took pity on me and—invoking her status as dean of the White House press corps—ended the briefing.

It took me months to dig out of that hole. I wasn't the first press secretary to be cut out of the loop when a military strike was underway; Larry Speakes, President Reagan's spokesman, denied that U.S. troops were going to invade Grenada, just hours before they landed on the beaches. In hindsight, I realize that I could have structured my guidance more carefully. I could have answered the first question about the original investigation—and then told reporters I would have nothing further to say until the president had made a decision. But the problem wasn't just that my guidance had been too specific—or that information I'd gotten from the NSC was wrong. The problem was that none of the people who knew about the strike even thought to find out whether I'd been asked about it. The daily briefing just wasn't on their radar screen.

There were other consequences to devaluing the press secretary's job. The relationship between any sitting president and the reporters who cover him has long been strained. But appointing a press secretary with less stature (and restricting the areas in the West Wing where reporters could go without an escort) was seen as an affirmatively hostile act, a statement by the White House that the press was less important and that access would be limited. Simmering resentments spilled over into public skirmishes, and both sides dug in. It would take several years—and a new press secretary—to repair the damage.

I know that neither Bill Clinton nor other members of the transition team foresaw the problems they would create with my compromised appointment. Maybe they should have, but they didn't. It wasn't in the president's interest any more than it was in mine. But they were trying to solve a different set of problems: problems of campaign promises and political correctness. Clinton needed visible women. And he wanted credit for appointing women to important posts, including the first to serve as the president's press secretary. But when all was said and done, some of those women, including me, didn't have the tools we needed to be effective in our jobs. And that made for some very difficult days.

Shortly before I left the White House, Larry King did a show featuring the White House correspondents from the major networks. At one point, he asked them about me. "Has Dee Dee Myers been effective?" he tossed out to the panel. "We'll never know," answered Brit Hume, then of ABC News. "She had an almost perfect temperament for the job . . . but she never really got the chance."

That wasn't the answer I would have hoped for when I started the journey. But after all I had been through, I thought it was fair, and I was grateful to Brit for his even-handedness. It also summed up how I felt: I never really got the chance. Don't get me wrong: I understand that in my case, gender cut both ways. I got the job because I was a woman. And I didn't quite get the job because I was a woman; I don't believe a man would have been appointed and given the same limitations. Still, I'm grateful for the opportunity. Despite the difficulties, it changed my life and created a wealth of opportunities that I wouldn't have otherwise had.

As I look back on it now, I understand that my experience is all too common. Like so many women in public life, I had a job with more responsibility than authority, and I struggled to make it work. Other women face the same broad range of obstacles: They're denied promotions, blocked from the informal networks that give rise to opportunity, judged by different standards. And together these obstacles keep women from being all that they can be. I share my story in the hope that it will help other women avoid some of the traps that I fell into—and to reassure them that they're not alone if they do. And I share my story because looking honestly at what happened to me—and why—has helped me to understand that some of the forces that shaped events were bigger than I. That's not to say I didn't make mistakes. I did. And if I had a chance to do it all again, there's plenty that I would change. But there's plenty that I couldn't change. Understanding that has allowed me to stop blaming myself for everything that went wrong—and start taking credit for some of the things that went right. And that's made all the difference in the world.

Chapter 2

Why Can't a Woman (Be More Like a Man)

"No book has yet been written in praise of a woman who let her husband and children starve or suffer while she invented even the most useful things, or wrote books, or expressed herself in art, or evolved philosophic systems."

—Anna Garlin Spencer,
American educator, feminist,
and Unitarian minister

As I sat in a conference room, somewhere in Naples, Italy, I could hardly believe what I was seeing: Laura Tyson, the chairman of the president's Council of Economic Advisors, couldn't get a word in edgewise—or even a seat at the table.

We had just arrived in Europe for the G7, the annual meeting of leaders from the world's big industrial nations, when the dollar took a precipitous tumble in international currency markets. The president gathered his economic and communications aides, as the world watched to see how he would respond.

There was a table in the center of the room, with additional chairs lining the walls, and as we filed in for the meeting, people assigned themselves seats based on seniority and role. The president; Lloyd Bentsen, the secretary of the treasury; and Bob Rubin, chair of the National Economic Council, grabbed chairs at the table, while several of their deputies and I headed for the outer ring. By the time Laura got into the room, all the seats at the table were filled—several by lower ranking men—and she had to settle for one of the cheap seats along the wall.

As options were being discussed, she tried to interject, only to be talked over. She tried again—and again, she was interrupted. Finally, she stood, her voice rising, and she started talking. As the men in the room looked on, a little startled, she kept talking, until the president recognized her and she was allowed to have her say.

Eventually, the president took Laura's advice—that he should say nothing—but that's really beside the point. Unlike the men in the room, she had to fight like crazy just to be part of the conversation.

Clearly, being in the room was not the same as being at the table. And it was hard to watch a woman with Laura's experience and credentials still struggling to be included. As we left the room, I sidled up to her. "I can't believe what just happened in there," I told her. "Unbelievable," she answered, slowly shaking her head, and we headed off to fight the next battle.

More than a decade has passed since the G7 met in Naples, and women have continued to gain ground in virtually every imaginable area of public life, from politics and business to science, academia, and sports. But in spite of the progress, women are still fighting their way in from the margins. There's still a double standard that holds women back. Male attitudes, ideas, interests, views, values, and voices are the norm. And since females don't necessarily share them, it's still too often seen as proof positive that they don't quite get it. Women are considered just a little less competent. Their problems are just that much less urgent. Their experience is not quite as valid.

The origin of this double standard isn't much of a mystery. For, oh, several millennia, women were confined to private life, where they raised children and managed domestic matters. Public life was the province of men, created by and for men. When women started moving into this traditionally male bastion, they had to take that world as they found it. From the earliest days, women succeeded by adopting the rituals of men, by going native. But women aren't men. So accepting the idea that they should behave like men—but don't—has created this sense that they're a cheaper model, that they're a Toyota to the male Lexus—same manufacturer, but without all the extra horsepower, fancy upholstery, and state-of-the-art electronics.

Women are caught in a double bind: expected to act like men—and punished for doing just that. According to a recent study by Catalyst, both men and women perceive women as better at "caretaking skills," like team building and encouraging others, while men are perceived as better at "taking charge" skills, like influencing superiors, solving problems, and making decisions. Empirical research shows that while both genders use a wide variety of leadership styles, well-worn expectations persist: When women in positions of authority conform to traditional female stereotypes, they are too often perceived as "too soft" to be effective. And when they defy those norms, they are considered "too tough," unnaturally masculine, out of sync. Damned if they do, damned if they don't.

The prevalence of these confining stereotypes is still an enormous barrier to women's advancement—and one that is widely recognized by women in leadership positions, even as it is often underestimated by others. Studies show that men tend to evaluate women leaders more harshly than do other women, which makes it harder still for them to break out of these boxes.

As the first woman to serve as White House press secretary, I definitely felt at times trapped by these competing expectations. I was supposed to be authoritative; after all, I was speaking for the President Of The United States, The Most Powerful Man On Earth. But at the same time, I had to be likable—a quality that's a bonus, not a requirement, for men in the same position. If there was a way to do both, on many days, it eluded me. At various points, as reporters predictably tested me, I tried getting tough. I'd pick apart their questions, challenge their assumptions, try to trip them up. It didn't work, even when I was right on the merits. Titles be damned, a thirty-one-year-old woman can rarely best an older, more experienced man—and win the exchange in any meaningful way. When I softened my approach, I often sounded tentative, which was like shoveling verbal chum into the shark tank; pretty soon the whole room would be trying to take a bite out of me. With more experience, or a few more rings around my trunk, I might have found a more effective middle ground. But the double bind would still have defined expectations for me in ways that none of my male predecessors had faced.

Of course, it isn't just the "firsts" who face these particular hurdles. Women in positions of leadership, in virtually every field, report similar experiences. Whether women are struggling to make a point in a private room—or being sized up on the public stage—the double standard persists.

Double Standard, Double Bind

During the transition, as Clinton worked to fill positions in his cabinet, a team of lawyers, media experts, and FBI agents vetted the various candidates, working to ensure that there were no skeletons in their proverbial closets. Along the way, researchers uncovered a variety of potential problems, some more serious than others, and in each case, a decision had to be made about how to proceed. In some cases, candidates were dropped from consideration; in others, they were allowed to go forward.

In the search to find the right woman to serve as attorney general, Warren Christopher, who was leading the transition, proposed Zoë Baird, a brilliant lawyer whose reputation was as impressive as her résumé, though she was just forty years old. Christopher had met Baird while he was negotiating the release of the American hostages in Iran and she had been part of his team. She later joined Christopher's firm, O'Melveny & Myers, before becoming general counsel to the insurance giant Aetna Life and Casualty. During the vetting process, Baird told transition officials that she and her husband had hired a Peruvian couple—she as a nanny, he as a driver and handyman—though both were in the country illegally. At the time, it was not illegal to hire workers without the proper documents, but Baird and her husband had also failed to pay their Social Security taxes, which was against the law. (After Clinton was elected, but before her nomination, Baird and her husband had hired a lawyer to help the couple obtain legal status and paid their back taxes and penalties.)

As transition officials debated how serious a problem this would be, one of the vetters, Ricki Seidman, was especially worried. The president-elect's nominee to become the highest law enforcement officer in the land had herself violated the law, and Ricki believed it would be a big deal. The story would be exacerbated, Ricki believed, by the fact that Baird was a woman (yes, a woman), that she had a nanny, and that even though she earned half a million dollars a year, she had failed to pay the proper taxes. When she asked me for my opinion, I said I wasn't sure it was disqualifying. But the fact that Ricki was so nervous made me nervous; I thought her instincts on these kinds of things were spot-on.

Others didn't see it that way. The nomination went forward, and when an avalanche of criticism made it clear that Baird couldn't be confirmed, her nomination was withdrawn. In the years since, I've often wondered: If Ricki had been a man, would things have gone down differently? Would she have been as sensitive to the implications of Baird's situation? And if so, would her opinion have carried more weight? Similarly, if Zoë Baird had been a man, would it have changed things? Would the public have been as incensed by her failure to pay Social Security taxes on her nanny? Or would they have simply assumed that it was the wife's responsibility? How much anger was generated by her failure to comply with a law that was, at the time, widely ignored—and how much by the fact that she was a powerful woman who left her children in the care of another woman?

The same kind of double standard applies to women running for elective office. Research shows that voters—both men and women—judge women candidates by different standards. *Cracking the Code*, a handbook for women candidates, warns them to be prepared. "Voters focus on a female candidate's performance under pressure, knowledge of issues, and personal presentation.

Voters are more judgmental about a female candidate's performance and less forgiving of her mistakes than they are of her male counterpart's. Voters ask: Can she stand up to her opponent? Can she think on her feet?" In other words, voters assume men are tough enough. But women have to prove it.

It's a fine line. Studies consistently find that if a woman is too assertive, it can hurt her chances of getting a job or advancing in her career. Simply disagreeing with men can sometimes get women into trouble. "I think [men] won't tolerate some things, unfortunately, from women," Kathleen Sebelius, the governor of Kansas, tells me. "And it's not fair, and it doesn't make any sense, but it's real. So either you can figure out that dynamic or you can not be as successful as you could."

As the first woman to run a truly serious campaign for president—meaning, one with a legitimate chance to win—Hillary Clinton has worked to strike just the right balance. In some ways, her challenges are as familiar as her black pantsuits: She's told friends that she worries about how to be strong without seeming strident. But in other ways, she's out there all by herself, trying to find balance on the highest of high wires, with massive expectations and little margin for error.

Even what to call her becomes a challenge. When she first entered public life, as the wife of the governor of Arkansas, she called herself "Hillary Rodham Clinton." But when voters fired her husband after his first term, she dropped the "Rodham." (He was subsequently elected again, then re-elected to three more consecutive terms, one assumes not entirely as a result of the name change.) When I first met her during the 1992 presidential campaign, she was introduced to me as "Hillary." And while most of us younger staffers referred to the candidate as "governor," everyone called his wife by her first name. When she became First Lady, I publicly referred to her as "Mrs. Clinton," but privately, she was still "Hillary." Ditto when she was elected to the Senate. Now she's running for president, not as "Senator Clinton" or "Mrs. Clinton" or even "Clinton," but as "Hillary." On the one hand, it's what people call her; it's breezy and familiar and warms that cool exterior just a touch. It also makes it easier to distinguish her from a former president with the same surname. But at the same time, it does undermine her authority ever so slightly. After all, "Hillary" sounds less formidable than "Senator McCain" or "Governor Romney" or "Mayor Giuliani."

(Throughout this book, I refer to her as "Hillary." That's how I came to know her, and that's how she now identifies herself. But should she be elected, I'll happily switch to "Madam President" before you can say "Hail to the Chief.")

Hillary's balancing act doesn't become any easier when the questions are not about nomenclature, but about policy. How, after all, can she convince voters she wants to play pat-a-cake with their toddlers—and nuke Iran? As Hillary has learned, that's one hell of a straddle. The risk—and often in her case, the reality—is that she falls short. "What I hear from young women to the old, what I call hard-line activists, is that she has become brittle," one prominent professional woman around Hillary's age told me. "And actually in many ways, it is almost like saying she has become male. They don't trust her. They don't know where her authentic being is anymore. And they just don't think she's real. And the one thing women have got to have—you've got to believe them. You've just got to believe they know who they are. That they have their own voice. And that it's real."

In fact, when Hillary showed a flash of the "real"—briefly tearing up near the end of the New Hampshire primary—women rallied to her defense, suggesting, perhaps, a growing weariness with rules that sometimes seem rigged. Barack Obama had come surging into the state in the wake of his impressive victory in Iowa, and virtually every poll confirmed that he would win comfortably in New Hampshire. But the political chattering class didn't just suggest that a second consecutive Obama win would cripple Hillary's campaign, they seemed to take delight in her demise; they were dancing at the edge of her political grave. As Hillary struggled to save her faltering campaign, she seemed worn down not just by the pace, but by the harsh and relentless criticism.

"How do you do it?" a woman asked her at a campaign stop.

"It's not easy," Hillary answered. "I couldn't do it if I didn't passionately believe it was the right thing to do." She was emotional but composed, as her eyes welled with tears that never quite spilled onto her cheeks. "This is very personal to me. It's not just political; it's not just public. I see what's happened," she continued, no doubt referring to seven years of Republican rule. "We have to reverse it."

For the next twenty-four hours, the moment played in near continuous loop on cable television and the Internet, as reporters and pundits debated whether the emotion was manufactured and how bad the damage might be. Even the normally sensible John Edwards suggested that if she couldn't survive the campaign without crying, she might make a weak commander in chief. Never mind that the last two presidents confessed to crying regularly—and were praised for their compassion. In a woman, it was a sign of weakness. Except that it wasn't. As voters watched it for themselves, women in particular seemed to conclude that her emotion was real—and that it was appropriate. Moreover, they felt her pain, and they decided to help end it.

The following day, women flocked to the polls. They made up a record 57 percent of the Democratic electorate in New Hampshire, and half—including many who hadn't previously supported her and many who might not stick with her—voted for her. Rather than losing by double digits, as the polls had suggested, she won; she found her "voice," as she said in her victory speech.

Too male. Not male enough. Too female. Not female enough. There are so many ways for women to lose at this game.

Judith McHale, the former CEO of Discovery Communications, hears the same kinds of complaints about Hillary Clinton, but she finds many of the arguments absurd. "Nothing drives me crazier than people who say, women who say, 'She would be so divisive.' And I say, 'Well, you know that would be horrible because we as a country right now are so united.'"

McHale chalks much of it up to a double standard. She recalls a conversation she once had with a woman who told her that Hillary made her uncomfortable. "And I said, 'Why do you feel that way?' And she said, 'Well you know she's an opportunist.' And I said, 'Well you know that is true. Certainly there is no one else in the U.S. Senate like that. Most politicians are really out there for the world, and she is the exception. She is the exception to all of that.'"

A VIRTUAL MUTE BUTTON

But if acting like a man can be a handicap, so can looking like a woman. And no one knows this better than Hillary. For more than a decade, her appearance has been the subject of an ongoing and robust national conversation. She should wear pants. No, skirts.

Her hair is too long. No, it's too short. Her gown is too conservative. No, it's too couture—and it makes her look fat! Once, when she showed up on the Senate floor wearing a low-cut camisole under a suit coat, you would have thought she had done the tango with Castro on the Capitol steps. Just the hint of cleavage—ohmigod, cleavage!—sent the chattering class into a swivet.

Similarly, when Katie Couric joined CBS as the first woman to serve as sole anchor of the evening news for a major broadcast network, the publicity department Photoshopped a picture of her so she would look a dress size or so thinner. "The emphasis on Katie's appearance—I hate it, it's so frustrating," says CNN's Campbell Brown, a friend of Couric's. "You don't hear the same kind of comments about male anchors. You just don't."

That said, the fashion police are increasingly turning their sights on men. John Kerry and John Edwards both got scorched for their fancy, and expensive, haircuts. But the conversation is almost always about the cost—$400!—or the circumstances—he had a stylist flown in from *Los Angeles*! It's almost never about whether it looks good or bad, whether it flatters the cheek bones or exposes the ankles. When it comes to women, everyone wants to be Joan Rivers on the red carpet at the Oscars.

From the time I joined the Clinton campaign, virtually every story written about me included observations about my hair, my earrings, my makeup, my clothes, my coat. Yes, my coat—my full-length black leather coat. I was from Los Angeles; at the start of the campaign, I didn't have a winter coat. It was December, and we were starting to spend a lot of time in New Hampshire; I needed a winter coat. I saw this leather number in the Eddie Bauer catalog. It was long (midcalf), waterproof, and had a zip-out down lining. I knew instantly that was the coat for me. And it remains the warmest coat I've ever owned. (Hillary

Clinton was so taken by my coat that she ordered the same one, a fact the press never seemed to notice—a small miracle, under the circumstances.) So the coat became part of my story.

Then there were the shoes. Again: I was from Los Angeles. I didn't own a pair of winter boots. Oh, I had some cowboy boots. It was the early 1990s; everyone had cowboy boots. It's just that I actually took mine to New Hampshire. In the winter. One day, as I was climbing out of the motorcade onto a slippery side street in Manchester, my feet came rocketing out from under me. Clinton caught my elbow, just before I went careening into the street. He looked at me, then at my cowboy boots. "You really are from California, aren't you?" he said, with a grin that let me know he was enjoying the moment a lot more than I was. Needless to say, I traded the cowboy boots for a pair of black Timberland snow boots.

There continued to be some focus on my footwear, as I somehow remained a little oblivious. In January of 1994, we were in Moscow for a presidential summit. Among other things, President Clinton and President Yeltsin would be signing a "de-targeting" agreement, each committing to point their nuclear missiles away from the other's cities. It was an important, symbolic step, and the two presidents were set to announce it at a press conference at the Kremlin. Since we would be playing by hosts' rules, the plan called for my Russian counterpart and me to call on reporters, as each president answered three or four questions. In other words, I would be on camera around the world for a considerable amount of time. We had a couple of other stops en route to the event, and the streets of Moscow were deep in gray, slushy snow. So before leaving the hotel, I slipped into a pair of snow boots. A few stops later, Mark Gearan, the White House communications director and my pal, took a look at my footwear. "You're not wearing those to the press conference, are you?" he asked, somewhat horrified. When I told him they were all I had with me, he grabbed his cell phone, tracked down someone from the advance staff and sent him to my room for a pair of black suede pumps. They arrived just before the press conference.

A few weeks later, I got a photo from the event. It's a head-to-toe shot of me, standing in front of a microphone, as Presidents Clinton and Yeltsin look on. The yellow Post-it Note is still attached: "Shoes by Mark."

And then there was the hair. Oh, the hair. I cringe now when I see the pictures. Blond. Blonder. Less blond but longer. Short again, with roots. Shorter still. Blonder. It was a march from bad to worse, back to merely awful. I knew people were talking. But only with distance and in hindsight can I appreciate exactly what they were saying. I wasn't a fashion icon.

It may seem counterintuitive, considering that I'm arguing *against* double standards, but one of my real regrets about my White House years was that I didn't do a better job with my appearance. Fairly or unfairly, the fact that I didn't get it quite right became a distraction—and given all the obstacles I was already facing, it was one I didn't need. It wasn't that I didn't try; I did. But fashion has never come naturally to me. I can't hear when I sing off key, and I'm never quite sure when my ensemble doesn't work. I wish I'd gotten help. When I came to Washington, just months past my thirty-first birthday, I'd never lived or worked in the capital, and with one or two exceptions, the only people I knew were those who had worked on the campaign. I didn't have another support network. I didn't have mentors, or even more seasoned girlfriends, who could have instructed me in the ways of Washington, helped me interpret the tribal rituals—or taken me shopping for shoes and suits. And I could have used that.

I recently discovered that the current White House has a full-time makeup artist on staff. The president is her primary client, but she's also available to the press secretary and any other member of the staff scheduled to appear on camera. It's a practical—and wise—nod to the power of moving pictures, and I can't help but wonder: Could I have pulled that off? Almost certainly not. First, there were fewer outlets—television, cable, Internet—hungry for video footage of the White House a dozen years ago. More importantly, it took a man—in this case, the president—to make makeup for events more important than driving to work but less important than the State of the Union smart, rather than vain.

But that was then. At an event during Inaugural Week, just days after I had arrived in Washington, I ran into a woman who had been appointed to a senior and visible job in the new administration. She was attractive and extremely well put together, and I complimented her on her suit. "I just spent $25,000 on new clothes," she confided to me. I was flabbergasted. That exceeded my entire net worth. I had never in my life known anyone who had spent $25,000 on new clothes in a few weeks. But the sartorial gap between us wasn't just about resources. There were clearly people in my financial situation who "got it." Like George Stephanopoulos. I remember walking into his transition office just days before we left Little Rock for Washington. There were some boxes leaning against the wall: long, wide, and oddly flat. "What's in the boxes?" I asked. "Four new suits from Barneys," he said. I'd never set foot in Barneys. George had managed to get himself to New York so that he'd be ready to stand behind the White House podium on Day One. Even if he wasn't a woman, he understood that the world would be watching.

Even now, virtually every time I appear on television, I get more feedback about how I look than about what I say. And that's just a reality that women face: People don't hear a word you say until they get over your hair. If you change your hair, people can't stop talking about it. And there's code. If people say, "Your hair looked great; did you get it cut (or let it grow)," that's good. It means they were generally satisfied with your appearance and could therefore move on to listening to what you had to say. But if they say, "I saw you on the *Today* show; you changed your hair," that's bad. It means they didn't like the way you looked and spent the entire segment thinking about what was wrong. In short: A bad hair day is a virtual mute button.

IN THE ROOM AND AT THE TABLE

When women aren't being judged by their appearance, they're often treated like they're invisible. Almost every professional woman I know—whether she works in law, business, medicine, academia, politics, or something else—has had this experience: You're sitting in a meeting in a room full of men. You make a suggestion, and no one responds. A few minutes later, a man says virtually the same thing, and everyone agrees that it's a great idea. What gives? To be sure, some women don't state their views forcefully enough. But there's no way that can account for the frequency with which this happens. In a recent study looking at various strategies women use to confront bias, one of the respondents suggested that when someone tries to "restate" one of their ideas, women have to confront them directly. "You've got to nip it in the bud," she said. "So you need to, with the right finesse, be able to go back to—let's say it's Joe Smith—and say, 'Joe, it's so great you thought my idea was right on target. I like the way you've reworded it, and you are exactly on the point I was on, and so'—to the collective audience—'what do you think about implementing my suggestion that Joe Smith just articulated so nicely?'"

Too often, women are expected to demonstrate abilities that men are assumed to have, to act like men and not act like men at the same time, to look the part without trying too hard. The expectations are often hard to define and harder to measure. Why can't a woman be more like a man? Call it the Eliza Doolittle syndrome. In spite of it, women are increasingly finding their way into rooms that had been previously closed to them. But when they do, they are still too often alone.

Not long after Katie Couric became anchor of the *CBS Evening News*, she was invited to the White House along with other network and cable anchors and the hosts of the Sunday morning political talk shows. Senior White House officials were set to brief them about a speech the president would be giving that evening, outlining a new course in Iraq. In her blog on the CBS Web site, Couric recounted that the surroundings were impressive, "even a little awe-inspiring." But the makeup of the room, she wrote, was "a little disconcerting as well." Save some female support staff near the doors, she was the only person in the room—from either the press or the administration—wearing a skirt.

I'd been in her pumps, so to speak. While I was White House press secretary, I participated in a number of similar briefings before major presidential events. I was often the only woman in the room, though sometimes Judy Woodruff, who was then hosting *Inside Politics* for CNN, was also there. But if there was a shortage of women, no one but me seemed to notice. And it wasn't necessarily because the reporters in the room had loftier topics on their minds. At one anchor lunch before the annual State of the Union address, the assembled journalists were riveted not by what was on the president's mind, but by what was on his plate. When waiters offered guests chicken, beef, and lamb on platters, French service, the president chose all three. And the press couldn't stop talking about it.

So what changes when women are in the room *and* at the table? Of course, that depends on the women, the room, and size of the table. Sometimes it doesn't change enough. Other times, just one determined woman with a point of view, a little power, and the willingness to shake it up can make a huge difference. Take the case of Dr. Bernadine Healy, an accomplished cardiologist and former director of the National Institutes of Health.

As crazy as it seems, it wasn't until the early 1990s that clinical health studies routinely included women. Until then, even studies that claimed to impart "scientific" wisdom about virtually all humans did so by ignoring half of those same-said humans. In the 1980s, for instance, a study examining risk factors associated with heart disease studied 15,000 men—and no women. Another exploring whether aspirin therapy could prevent heart attacks focused on 22,000 men—and no women. And one that asked whether estrogen protects against heart disease studied—you guessed it—no women! In other words, almost everything that was known about women's biology was learned by studying men. I can assure you that the people who devised this system were not women, since most of the women I know are quite certain that their bodies are not just like men's, only with different hardware.

For years, women's groups—and especially, the small but growing number of women in Congress—had been demanding that the federal government take women's health issues more seriously. The treatment of heart disease in women and of women's cancers, and the role of hormonal changes and therapies, had been understudied and underfunded, they argued. And yet, precious little attention had been paid to this particular problem. It was the public policy version of "It's all in your head, dear."

All that changed when Healy became the first woman to lead the National Institutes of Health in 1991. Within months, she made it clear that women could no longer be excluded from the agency's clinical trials. Existing NIH policy already said as much, but the rules had never been enforced. "So when I got there, I said, 'Guess what? We're going to enforce it,'" she told me one morning in her office at *U.S. News & World Report*, where she is now a columnist. And they did.

Soon after, Healy launched the Women's Health Initiative, a study that would follow 150,000 American women for more than fourteen years to determine the causes of disability and death in women over fifty. The study would look at whether hormone replacement therapy, vitamin supplements, and other changes in diet could help prevent heart disease, breast and colon cancer, and bone loss in post-menopausal women, while at the same time developing an enormous database on issues related to women's health, and has provided invaluable information, such as the link between hormone replacement therapy and breast cancer. It was an important step in a long-overdue effort to close the fact gap between what was known about men and what was known about women.

It wouldn't have happened without Healy—and the women in Congress—who understood that the lack of attention to women's health issues was inexcusable because it was dangerous. For example, heart disease is the number one killer of both men and women. But until recently, *all* the research into the disease was conducted on men. As a result, doctors and their patients believed that men and women experienced heart attacks in the same way—massive chest pains. To be sure, some women do have chest pains. But they are more likely to experience other symptoms: shortness of breath; flulike nausea, clamminess, and cold sweats; and pain in areas other than the chest, such as the shoulders, neck, or jaw. But since women don't always recognize the symptoms, they are less likely than men to believe they're having a heart attack—and more likely to delay treatment. No doubt, thousands and thousands of women have died or been disabled unnecessarily as a result of a culture that paid less attention to women's health than to men's.

The Smart Thing

So what might we expect from a world without double standards? In a word: equality. But equality doesn't mean that men and women are the same. It doesn't mean that women have to try harder to act like men, think like men, look like men. "Because if women buy into that, then everything else becomes men are the normative standard of behavior in the world outside of the home," Healy explains. "Outside of maternal behavior, men are the standard. And that has ramifications that I think have filtered into education, filtered into keeping women down. Because women can never play that game, any more than you would ever expect a man to do the same. So part of even the women's health movement, I think, was recognizing that women are different."

Different. But equal. That doesn't mean that every man should be expected to behave one way, nor every woman another. Rather, it means that women's ideas and opinions and experiences should be taken as seriously as men's—regardless of whether they conform to traditional stereotypes. And it's not just about doing the "right" thing; it's about doing the smart thing.

Increasingly, women are the engine driving economic growth worldwide. Since 1980, women have taken two jobs for every one filled by a man. And their influx into the workplace has contributed more to global economic growth than either new technology or the new giants, India and China. What's more, when you add the value of housework, child-rearing, and other domestic chores, women probably account for more than half the world's output. But women aren't just workers. They are increasingly important as managers, consumers, investors, entrepreneurs, and directors.

According to a recent study by Catalyst, Fortune 500 companies with the highest representation of women on their boards performed better financially—significantly better. When compared to companies with the fewest women in the board room, those with the most saw a 53 percent higher return on equity, a 42 percent higher return on sales, and a 66 percent higher return on invested capital. Moreover, the findings were consistent across most industries, from consumer goods to information technology.

Why? Among other things, companies that are willing to look beyond the traditional (read: white and male) labor pool are finding a large and growing reservoir of talent. Women now earn 60 percent of bachelor's and master's degrees, and nearly half of all doctorates, law degrees, and medical degrees. As the world changes, women are particularly well qualified to contribute in a marketplace that values brains over brawn—and new ideas above all.

Women already make the vast majority of purchasing decisions, and their financial power continues to grow. Women account for nearly half the workforce, and between 1990 and 2003 their median income grew 25.6 percent, compared with 8.1 percent growth for men. According to *BusinessWeek* and Gallup, women will control some $12 million—more than half of America's wealth—by the year 2010.

In 2003, De Beers, the diamond industry conglomerate, launched a new ad campaign that recognized these changes.

"The Left Hand Rocks the Cradle; The Right Hand Rules the World," read the caption, below a picture of a woman's right hand sporting a bit of bling on the third finger. To be sure, the right-hand ring is not new; it used to be called a "cocktail ring." What is new is that women are increasingly buying them—and other pieces of jewelry—for themselves. While arguably more a reaction to this growing trend than the cause, the De Beers ad nonetheless reflects a changing marketplace.

A decade earlier, the chief strategist at Goldman Sachs in Tokyo—a woman—recognized this trend and devised a basket of 115 Japanese companies that she believed would benefit from the changing needs and increased purchasing power of women. It included everything from financial services to online retailing, beauty, clothing, and prepared foods. Over the past decade, the value of shares in this basket has increased 96 percent—while the value of the Tokyo stock market has risen only 13 percent.

Not only are women a wise investment, they are also wise investors. Numerous studies suggest that—contrary to conventional wisdom—women "consistently achieve higher financial returns than men do." They tend to do their homework, buy and sell based on information and thoughtful consideration, and keep their portfolios balanced. Men, on the other hand, are more likely to buy on a tip, overinvest in "hot" stocks, hold onto losing investments too long, and churn their accounts, which reduces return.

Women aren't just changing the marketplace, they're also changing politics. As voters, women increasingly vote independently of their husbands, fathers, and brothers. In every presidential election since 1980, there has been a gender gap. And in each of the last three, men and women, voting alone, would have elected different candidates. In addition, studies show that as elected leaders, women feel a special responsibility to represent women. True, ideology is generally more important than gender when it comes to politics and policy; not all women think and feel the same about important issues. Still, women are more likely to introduce and support legislation benefiting women, children, and families, regardless of party. Experts also say that voters tend to believe women are "better listeners, more honest and can work across party lines." So at a time when progress on important issues has been stalled by intense partisanship, the increased presence of women can help reassure citizens—and break the logjam.

Nobel Prize laureate economist Amartya Sen believes that "nothing, arguably, is as important in the political economy of development as an adequate recognition of political, economic, and social participation, and leadership of women." Empowering women, he argues, not only improves their well-being (and that of their children), it also leads to other, broader social changes. For example, educating girls yields higher return than educating boys in many countries. It does more to lower fertility and infant mortality rates, and improves agricultural productivity.

Without a doubt, the increased presence and power of women in public life has generated enormous, positive change. But getting to a place beyond double standards, where equality is not a slogan but a way of life, will demand more. It begins with acknowledging that men and women are different. And it embraces the idea that because they are different, women will bring with them a different mix of experience, values, and points of view. That, in turn, will expand the range of what's acceptable—and what's possible. It won't be easy; if it were, it would already have been done. But it's in our economic, social, and political interest to create a world that's freer and fairer. Where women have more power—and are allowed to use it. Where everyone is judged by their performance—and their potential. Where double standards are only a distant memory.

Chapter 3

Biology, Ideology, and Difference

"What do you call a person who believes boys and girls are the same? Childless."

—Steven Pinker

I have two young children, a girl and a boy, and I'm endlessly fascinated by the ways they are alike—and the astonishing ways in which they are different. Unlike me, my daughter is kind of a girly-girl. From the time she was very young, she spent a lot of time feeding her dolls and stuffed animals, caring for them when they were sick, and putting them to bed. For years, she left a trail of little toys tucked under baby blankets, washcloths, and scraps of tissue wherever she went. My son, who is three and a half years younger and was born into a house heavily stocked with "girl" toys, had a totally different reaction. It would never occur to him to put a stuffed animal to bed; he's far too busy trying to slay it.

Recently, my children were playing with their young cousin, a girl. I was only half paying attention, as they all seemed happily engaged in some iteration of "house." Suddenly, they started arguing. "There's no intruder in this game!" the girls insisted.

"If there's no intruder, who am I going to shoot?" my son wanted to know. The girls held their ground; they had no interest in allowing a violent confrontation to disrupt the tranquil rituals of their imagined domestic life. But unless my son had a mission, a bad guy to kill, a family to protect, he had no interest. So he quit. And so it goes, day after day, as I observe the different interests, attitudes, and goals of my children, in their relationships and in their play, in how they see the world and how they respond to it. I feel as if my life is a science experiment, and so far, the evidence overwhelmingly supports one conclusion: Boys and girls are born different.

Oh, I know, my stories—and I have *hundreds* of them—may be easily dismissed as a bit cliché, a playground version of Mars and Venus. But when I compare notes with the parents of other young children, they virtually all tell some version of the same tale. Their girls tend to talk early, play cooperatively, and develop a mysterious love of princesses at around three. (How do they even know about princesses?) Their boys will turn any object they find into a weapon. And no one ever says, "Gee, I think it's just the way we're raising them."

From as far back as I can remember, I knew that—duh—boys and girls were different. They looked different. They dressed different. And they were treated different. I played volleyball in high school, and the boys' basketball team *always* got first call on the big gym with the springy wooden floor when our practice schedules overlapped; we had to make due in the crappy, tile-floored, shin splints–inducing "girls' gym."

When I was younger, I didn't give much thought to *why* they were different. But by the time I got to college in the fall of 1979, I was pretty sure it had more to do with nurture than nature. Lopsided expectations, differential treatment, and traditional notions about who should and could accomplish what did more to shape gender roles than genes, I thought. And a lot of those ideas were the accepted wisdom not just in the classroom, but in the dorms and the dining hall, at student council meetings and during late nights in the production offices of the campus newspaper.

Over the years, however, scientific discovery and my own experience have reshaped my thinking—and changed my mind. Clearly, culture affects behavior in profound and complex ways. But I no longer believe that gender roles are arbitrary, as susceptible to change as hemlines and hairstyles. New tools have allowed scientists to find structural, chemical, genetic, hormonal, and functional differences in male and female brains. And those differences affect the way men and women process language, solve problems, and remember emotional events. They shape our responses to everything from stress, to love, to the funny pages. And it's increasingly clear that the answer to that age-old question isn't nature *or* nurture: It's both.

In some circles, the idea that nature accounts for any of the differences between the sexes is considered totally un-PC, totally unfeminist. Still, I think of myself as a feminist. When I travel around the country giving speeches, I often ask people to raise their hands if they also consider themselves feminists. *Way* more often than not, a solid majority—regardless of age—practically shudders at the thought; they'd rather post pictures of themselves in their underwear on the Internet than claim the label. It's not that they're all throwbacks to an Ozzie and Harriet era, when women's influence rarely extended beyond the end of the driveway. These same people believe that men and women should have the same opportunities, that they should be treated equally under the law, and that they should receive equal pay for the same work.

And those beliefs often transcend ideology. Danielle Crittenden, a conservative writer and social critic (no feminist, she!) believes that most women want pretty much the same things. In addition to husbands and children, "many of us will want to pursue interests outside of our families, interests that will vary from woman to woman, depending upon her ambition and talent. Some women will be content with work or involvements that can be squeezed in around their commitments at home; some women will want or need to work at a job, either full- or part-time. Other women will be more ambitious—they may want to be surgeons or corporate executives or lawyers or artists." While Crittenden warns that managing job and family will be difficult, she never questions the right of women to have and make these choices. I can almost hear the cosmo glasses clinking.

So what gives? Most of the feminist movement's original ideas have become mainstream, and even women who eschew the F-word take the gains for granted. Recently, a female professional fighter scoffed at the idea that her brand of mixed martial arts battles between women was not ready for prime time. "That is so ridiculous," Lisa King told *Time* magazine. "I'm not a women's lib person or anything, but we're doing everything else, why not this?"

So if that's not "women's lib," what is? What is it about the movement itself that feels like so many manicured nails on the blackboard? To a large degree, I think it's the idea that "equal" means "the same." The hardcore ideologues on the left flank of feminist thinking insist that gender roles are not just artificial, but designed to hold women back. So until they're destroyed, women will continue to be victims of a patriarchy, virtual slaves in their own homes. But most women—even many who call themselves feminists—don't want to be "the same" as men. Nor do they believe that equality demands it. That's not to say that questions about what equality means or how it might be achieved have been resolved. Far from it. But the vision of a one-size-fits-all world simply denies too many women's experience—and their aspirations.

Deconstructing Larry Summers

Saying there are differences is one thing. But suggesting that those differences might affect everything from aptitude to interest is like dancing in a minefield: A single misstep can have serious consequences. Just ask Larry Summers.

In January of 2005, Summers, who was then president of Harvard, gave a talk about why only about 20 percent of tenured professors in science and engineering at the country's top universities were women. He began by saying he wanted to be provocative—and by that measure, he was wildly successful. But whether he shed light on the question, or just heat, is less clear.

In typical Summers fashion, his speech strode confidently through an interesting analysis of the issue, discussing a series of current theories. And then he offered his opinion: that "the largest phenomenon, by far, is the general clash between people's legitimate family desires and employers' current desire for high power and high intensity; that in the special case of science and engineering, there are issues of intrinsic aptitude, and particularly of the variability of aptitude; and that those considerations are reinforced by what are lesser factors involving socialization and continuing discrimination."

Well, not all reasons are created equal. So while there's broad consensus that the challenge of balancing work and family, socialization, and ongoing discrimination all hold women back, the idea that they simply have less aptitude for hard science and engineering is, *ahem*, controversial. Even though Summers acknowledged that his conclusions might be "all wrong"—and he challenged those with different views to contradict him—the damage was done; his speech sliced head-on into a raw nerve, generating a seemingly endless, energized and withering response.

"Here's the president of Harvard who simply doesn't understand the impact [that attitude] has on my life!" said Sally Ride, the nation's first woman astronaut. "It's personal. That's why it got such an outpouring."

When I first heard about Summers's speech, I was baffled. What in the world was he thinking, I wondered? I'd known Summers from our days in the Clinton administration, and I'd always liked and respected him. Yet in spite of the good advice he consistently gave the president, he could be more than a little tone-deaf when it came to his own words. This seemed like a case in point.

But what, exactly, was his sin? The news coverage focused exclusively on his comments about "intrinsic aptitude"—and the furious response it generated. But surely there must be more to it, I thought, so I read the speech. And then I read it again. And there's no getting away from it: Summers said that men, by their essential nature, are more likely than women to excel at hard science and engineering. He didn't raise it as one of a number of theories that needed further examination; he offered it as a conclusion—even though it had yet to be "proved" scientifically. What's more, he demoted the role of ongoing discrimination, calling it a "lesser factor." And surprise, surprise, many of the women on the receiving end of that "lesser factor" went postal. He had crossed that sometimes-blurry line from inquiry into offense.

At the time, his relationship with many of the women on campus was already strained. A friend of mine who then worked at Harvard told me that women often complained that Summers could be rude and dismissive. Worse, they were angry that fewer women on the Arts and Science faculty were offered tenured positions during every year that he was president. In the end, the reaction to his speech, coupled with a faculty revolt over his management style, was more than even his high-profile campaign to apologize and make amends could overcome.

To some degree, I suspect he was showing off. He is, without question, a brilliant man, and he's not above taking his brain out for a spin when the other kids are watching. But he went too far. If he'd stuck to the evidence, marshalling current research to support—and contradict—each of the various theories, he would probably still be president of Harvard.

As Dr. Shirley Tilghman, the popular, first female president of Princeton and an accomplished scientist, told me later, "Larry put his foot in it in part because he was offering opinions about something he did not know enough about to really be offering opinions. It is a kind of third-rail issue. And when you're talking about a third-rail issue, you'd better know your data, and know what the situation is."

So Larry Summers jumped the gun, and he paid a heavy price. But there *are* numerous and complex differences between the genders. And I started to wonder: Do any of them really affect women's aptitude for certain kinds of careers? If women are better at reading other people's emotions, for example, might it also be true that men are better at quantum mechanics? I decided to look for answers.

To begin with, I found out that research shows there is no gender difference in general intelligence. And in areas where there are differences, some favor men and others favor women. Men are better at mathematical problem solving; women are better at mathematical calculation. Men are better at mentally rotating shapes; women are better at visual memory. Maybe this helps explain why more men are engineers but more women are accountants. In other instances, men and women use different processes to achieve the same result. For example, when solving certain problems like navigating a route, women tend to use landmarks ("go three blocks until you see the Exxon station on your left; then turn right past the park"), while men prefer geometry ("go a quarter of a mile north to a traffic circle; go 270 degrees around it, until you're heading west"). And maybe this helps explain why some women need to turn the map around when they're reading it, a habit that some men find baffling.

I also found out that there is a significant difference in "variation" between men and women. What does this mean? When you measure various qualities of both men and women—height, weight, intelligence, likelihood to become criminals, aptitude in science and math—men tend to differ more from each other than women do. For example, if you measured the height of 1,000 men, and plotted each one on a graph, you would end up with a bell curve. The largest group of men would fall somewhere in the middle, around five feet ten inches, give or take a couple of inches. And the farther you moved in either direction, the fewer points you would find on the curve; in other words, there are a lot more men who are five-eight than six-nine so the curve flattens out along both tails. If you did the same thing with women—measured 1,000 women and plotted their heights on a graph—you'd find that women are, on average, shorter than men. But that's not the point. There is a smaller range of heights among women; there is less variation. So the tails on both ends of the bell curve are steeper.

Similarly, if you measured the IQs of any number of men and women and plotted them on a graph, you would find that men and women are exactly the same in terms of average intelligence, and most cluster around the middle of the curve. But once again, men show more variation; there are more men at both extremes of the curve; there are more boneheads and more geniuses.

So what happens when you compare men and women, if you plot the two curves on the same page? When measuring height, you'll see that not all men are taller than all women. Some women are five-eleven, while some men are five-four. But as you go farther out along either tail, the number of men gets proportionally larger and larger than the number of women. At five-ten, there are thirty men for every woman, but at six-two, there are two thousand men for every woman.

When you look at intelligence, the differences are less extreme, but the pattern is the same: There are increasingly more boys to girls, the farther you move along the spectrum. On one end of the continuum, boys are much more likely than girls to be diagnosed with learning disabilities, mental retardation, or autism. On the other end, more men tend to have extreme high IQs, and more boys score perfect 800s on the math portion of the Scholastic Aptitude Test. But again, the question is, why? In the 1970s, thirteen times more boys than girls scored 700 or higher on the SAT. Twenty-five years later, that number is substantially smaller. The fact that the ratio has changed—and so quickly—seems to suggest it's affected by factors such as the way the test is written, or how much girls are encouraged in math and science in high school, which are unrelated to innate differences. And almost certainly, the numbers will continue to change.

THE 1 PERCENT DIFFERENCE

Genetically speaking, men and women are incredibly similar; 99 percent of our genetic material is the same. *Hmmm*, I thought. *Only a 1 percent difference; that doesn't seem like very much.* But when I read that chimpanzees differ from humans by just 1.5 percent, that small difference got pretty darn big.

"We all recite the mantra that we are 99 percent identical and take political comfort in it," says David Page, the MIT biologist who mapped the Y, or male, chromosome. "But the reality is that the genetic difference between males and females absolutely dwarfs all other differences in the human genome."

Increasingly, scientists are finding that gender begins in the brain. According to Sandra Witelson, the neuroscientist famous for her study of Albert Einstein's brain, "the brain is a sex organ." And increasingly, new tools—like PET scans and MRIs—are helping scientists understand how it works.

In her fascinating book *The Female Brain*, Dr. Louann Brizendine explains that male and female brains are indistinguishable in the weeks following conception. Then, at about eight weeks, the male baby gets a dose of testosterone, which literally begins killing off cells in the communication, observation, and emotion processing centers of the brain—and growing cells in the sex and aggression centers. It's sort of like "Wow!" and "Duh!" all at the same time.

Male brains are roughly 9 percent larger than female brains, even when you control for the fact that men are generally larger. Back in the day when bigger brains were thought to be better brains, the difference was "proof" that women weren't as smart. But it turns out that women have the same number of neurons; they're just packed into a smaller space. And in the areas of the brain that control language and hearing, for example, women have on average 11 percent more neurons than men.

In addition, the connection between the right and left halves of women's brains is larger, and each of the hemispheres less specialized. As a result, a stroke on the left side that might leave a man barely able to speak can be much less debilitating in a woman, since she uses both sides of her brain for language. While scientists are still investigating the effects of this more connected, more integrated female brain, some speculate that it may allow women to process information from a number of sources simultaneously. And this, in turn, may help explain not only women's sense of intuition, but also their ability to multitask. Women are, in the words of Mary Catherine Bateson, a cultural anthropologist and former dean of Amherst College, "peripheral visionaries," able to keep track of several ideas, tasks, or children simultaneously.

Throughout our lives, these differences manifest themselves in a variety of ways. In childhood, boys' play generally tends toward the rough-and-tumble. Like my son and his playmates, boys chase, push, and tackle each other, put Play-Doh in each other's hair, and turn any available object into a weapon to shoot, stab, or vaporize each other. It's not that they're more active; girls are just as active when there are other kinds of toys, like jump ropes or trampolines. But boys are more competitive—according to one study, an astonishing fifty times more competitive—as they seek to climb to the top of the pecking order. Girls, by contrast, are more cooperative, and their games often center around nurturing. They're much more likely to seek consensus and be more concerned with fairness, rather than competition. Studies show that they take turns twenty times more often than boys. The relationship—not winning—is the goal.

Simon Baron-Cohen, a professor of psychology and psychiatry at Cambridge University in England who has studied male and female brains extensively, describes it a little differently. He believes that female brains are wired to empathize, while male brains are designed to understand and build systems. In order to rule out (or at least severely limit) the influence of socialization, he conducted one study on infants who were just twenty-four hours old. When the babies were shown a mechanical mobile and a human face, the boys looked longer at the mobile, while girls preferred the face. Okay, I've had two babies. And while I was endlessly fascinated by every flutter of their newborn eyelashes, I'm not sure I could have ever really figured out what they were watching. Thankfully, the science didn't stop there.

Another study showed that year-old boys preferred a video of a car, while girls the same age preferred to watch a talking head, even with the sound turned off. That tracked with my experience. When my daughter was around one, my aunt Jane sent her a video called *Baby Faces*, a montage of different babies registering different expressions set to music. She loved it; it was the only one that would hold her interest for more than five minutes. But when my son came along a few years later, he had no interest in it. In fairness, I never showed him a video of a car; maybe I should have.

Baron-Cohen also measured testosterone levels in amniotic

fluid from prenatal tests, and later evaluated the children to see if there were differences in their social development. At each of three stages—twelve months, twenty-four months, and four years—the children with the lower fetal testosterone made more eye contact and had larger vocabularies. And over time, they continued to display better social skills, better language and communication skills, and broader interests.

It's not surprising, then, that girls and women tend to do better on tests that measure the ability to recognize emotion. In one test, participants were shown photographs of faces—but only the portion just around the eyes. They were then given a list of four words, like "angry" or "sad," and asked to pick the one that best described what the people in the pictures were thinking or feeling. The women scored higher. I can't say I'm shocked. More than once, I've been in a meeting or at a social event with a group of women and one or two men. Eventually, one of the men always seems to get that when-did-you-start-speaking-Swahili look in his eyes; he simply loses the thread of a conversation that requires a finely honed ability to navigate the emotional subtext. To be honest, I can sometimes get lost myself, while some men are right there, every step of the way. But most aren't. To their credit, men tend to be better at mentally rotating three-dimensional objects in their minds; they can imagine what the Buick might look like if it was suspended in space and rotated 270 degrees. Important, I'm sure. But unlikely to help them tell the difference between tired and really mad.

Not only are women (and girls) better at reading other people's feelings, they worry more about hurting them. As few years ago, ABC News correspondent John Stossel conducted a test with the help of a researcher from the University of California. Stossel's crew made lemonade, but rather than sweetening it with sugar, they loaded it with salt. They then gave it to groups of boys and girls, and filmed their responses.

"Eech!" sputtered two boys named Aaron and Jacob.

"It needs sugar," insisted Raja. "It tastes terrible."

But the girls reacted quite differently. "It's good," said Morgan.

Only when pushed did the girls tell the truth. "I didn't want to make anyone feel bad that they made this so sour," Asha explained.

The areas of men's brains that control action and aggression are predictably larger. And—are you sitting down?—men devote two and a half times the brain space to their sex drives! So while most women think about sex a couple of times a day, men's minds are stopping by the adult cinema once a minute. And what happens in the brain when someone falls in love? Again, that depends on whether you're a man or a woman. MRI scans reveal that in both genders, romantic feelings generate activity in the areas of the brain associated with energy and elation. But women showed more activity in the areas linked to reward, emotion, and attention. And men's brains showed activity in visual processing areas, including one associated with sexual arousal. Which means even more mental trips to the adult cinema.

What about humor? A few years out of college, my friend Sue and I went to see Jay Leno, who was performing his stand-up routine in a local theater. His show included a riff about gender. Forget everything you've heard, he said. The fundamental difference between men and women is that men like the Three Stooges and women hate them. *He's talking to me,* I thought! *I* hate *the Three Stooges.* Over the intervening twenty-plus years, I've occasionally met women who enjoyed watching Curly, Larry, and Moe bop each other on the head, but most women I know find them insufferable. Well, it turns out there may be biological reasons. In a recent study, researchers gave a group of men and women a series of cartoons and asked them to rate how funny they were. Using an MRI, researchers found that when women laughed or smiled, there was a burst of activity in the areas of the brain that govern language, sort through complex ideas, and monitor rewards. Men used a different, less sophisticated process to decide whether the cartoons were funny. According to Dr. Allan Reiss, the chief researcher, this helps explain why men laugh at flatulence jokes and slapstick. "It doesn't take a lot of analytical machinery to think someone getting poked in the eye is funny."

Men and women also experience relationships differently. Throughout their lives, women seek friendships based on intimacy and understanding. But men's friendships tend to be based more on shared interests; guys like to do stuff together. Surely, this helps explain the different ways men and women communicate. "For women, as for girls, intimacy is the fabric of relationships, and talk is the thread from which it is woven," writes Deborah Tannen. Just as girls create and maintain relationships by exchanging secrets, women view conversation as the "cornerstone of friendship." It is not so much what gets talked about; it's the talking itself that is most important.

"If you told a woman that you had just returned from a trip to the surface of the Moon, she would show her interest by asking who you had gone with," a man named Howard told the BBC recently.

Boys can form bonds that are as intense as girls, but you don't have to be a rocket scientist (or a professor of linguistics) to know that it's less about sharing thoughts and feelings than about sharing activities. Men can spend four hours on the golf course with their "close" friends and come home without a shred of information. And when they do talk, it doesn't look at all like a conversation among women. Tannen studied a series of videotapes of children and adults talking with their same-sex best friends. She found that "at every age, the girls and women faced each other directly, their eyes anchored on each other's faces. At every age, the boys and men sat at angles to each other"—like they were driving in a car, staring out the windshield. They looked around the room, periodically glancing at each other. They were listening—just not in the same ways that women listen.

According to studies, men's self-esteem derives more from their ability to maintain independence from others while women's self-esteem is maintained, in part, by the ability to sustain intimate relationships. And again, this explains a lot, including why men won't ask for directions. The satisfaction they get from getting to their destination without help simply outstrips the frustration they feel after spending all that extra time on the road.

My husband likes to go to the store, any store. Which is extremely handy. But what I think he likes best is finding stuff without help. He'd rather comb the aisles of an unfamiliar hardware store than (in his view) admit defeat and ask the guy in the red vest. Which drives me crazy. But now that I (sort of) understand his point of view, I don't (always) feel the urge to try to change him. He brings home the things we need—without my help (or what he considers my excessive questions about the relative merits of product A versus product B)—and we're both happy.

HYSTERIA AND HISTORY

So while there's a lot of evidence that men and women are innately different, I understand why some people, women in particular, are often reluctant to embrace those differences. First, the evidence can be confusing; each new discovery seems to raise as many questions as it answers. In addition, "science" has been used to define women—and their biology—in ways that have denied their competence, autonomy, and power for thousands of years. As a result, a lot of women are understandably nervous about letting the latest theories limit people's thinking—or their options.

"For most of this century, women were kept out of the legal profession, and they were kept out because it was said—by men—that women were too kind, too gentle to be good lawyers," says Sandra Day O'Connor, the first woman to serve as a justice on the United States Supreme Court. "To be a lawyer, you had to be tough-minded and rough-hewn and women were just not cut out for that kind of work."

As far back as ancient Greece, "hysteria" was understood as a medical condition particular to women. Plato described the womb as "animal within an animal," moving throughout the body and essentially wreaking havoc. Hippocrates, the so-called Father of Medicine, taught that it was the irregular movement of blood from the uterus to the brain that caused women to go bonkers. During the Victorian era, the diagnosis of hysteria was widespread, and the list of symptoms was long: "faintness, nervousness, insomnia, fluid retention, heaviness in the abdomen, irritability, loss of appetite for food or sex, and a 'tendency to cause trouble.'" The recognized treatment was "pelvic massage"—or the manual stimulation of the woman's genitals by her doctor to the point of "hysterical paroxysm," better known as orgasm. Ironically, the ideal woman of the era was considered "passionless," uninterested in sex except to have babies. But the causes—and cures—of hysteria were all about sex.

Given the number of afflicted women, the requisite frequency of treatment, and the fact that the desired result—under these somewhat less than romantic conditions—often took hours, the poor doctors must have been exhausted. But fear not, help was on the way. By the late 1800s, "hydrotherapy" devices started popping up at popular bathing resorts in Europe and the United States, followed by mechanical vibrators, reducing treatment from hours to mere minutes. By the turn of the century, as electricity was becoming increasingly available in homes, plug-in vibrators became hugely popular home appliances—widely available nearly a decade before either the vacuum cleaner or the electric iron. The 1918 Sears catalog carried a portable model with attachments that was billed as "very useful and satisfactory for home service."

Eventually, the diagnosis of hysteria declined, and it is no longer a recognized illness. But the idea that a woman's competence is totally controlled by the rhythms of her body has been harder to shake.

When female race car driver Danica Patrick considered switching from the Indy League to NASCAR, fellow driver Ed Carpenter took note. "I think Danica is pretty aggressive in our cars—especially if you catch her at the right time of month," he said.

And then there are those at the other end of the spectrum. I have a friend who is a psychologist, and she believes that most behavior differences are the result of nurture. For example, she says studies show that mothers use significantly more words when communicating with their daughters than with their sons, and this helps to explain the common observation that girls are more verbal. So I asked myself, did I talk more to my daughter at various developmental stages than to my son? And I realized the answer was probably yes—but for a reason. When my daughter was a baby, she'd lie on the changing table and look me in the eye. And even when she couldn't talk, she'd coo and gurgle at me, a full participant in the diapering, dressing, or bathing at hand. My son, on the other hand, was a whirling dervish. He was always reaching for a bottle of lotion—or worse, a dirty diaper. He was constantly trying to flip over to get a better view of the toy lying on the floor or the bird flying past the window. He wasn't much interested in the tasks at hand, and so my primary goal became not teaching him to say "Mama" but saving us both a trip to the emergency room.

As much as I hate to admit it, there is scant evidence that there has ever been a society that was truly matriarchal, where women ruled. What's more, sex roles have been stunningly consistent across time and culture. In every known civilization, in every period, in every spot on the planet, women have been primarily responsible for raising the children while men have ventured out into the world. Coincidence? I think not. Similar patterns are also found in most mammals. Males tend to be more aggressive, even in the preference for rough-and-tumble play. Females spend more time with their babies, and their play tends to be more nurturing. In fact, when infant monkeys are given a choice of human toys, males prefer trucks; females, dolls.

Some societies have tried to neutralize traditional gender roles, most famously on the Israeli kibbutz. In their socialist zeal to make sure everyone was treated the same, an equal number of men were assigned to care for children, and an equal number of women to repair tractors. Within a generation, however, both men and women gravitated back toward more traditional roles.

A lot of gender differences appear in early childhood, which suggests that they're not entirely socially constructed. And in instances where biological boys have been raised as girls, they end up feeling like, well, boys. In one famous case, a boy lost his penis in a botched circumcision. (His twin brother was spared the same fate.) On advice of "experts," his parents decided to castrate him, give him hormones and raise him as a girl. He was never told. But by the time he was fourteen, he was depressed and unhappy. Finally, his father told him the truth. After more surgery, he became a boy again. He grew up, married a woman, and adopted two children.

Acknowledging innate differences is not a matter of taking sides; rather, it's a key to leveling the playing field. If men and women are different, if their brains sometimes work in different ways and cause them to experience the world differently, then expecting women to act like men puts them at a substantial—and permanent—disadvantage.

"What if the communication center is bigger in one brain than in the other?" asks Dr. Brizendine in *The Female Brain*. "What if the emotional memory center is bigger in one than in the other? What if one brain develops a greater ability to read cues in people than does the other? In this case, you would have a person whose reality dictated that communication, connection, emotional sensitivity, and responsiveness were the primary values. This person would prize these qualities above all others and be baffled by a person with a brain that didn't grasp the importance of these qualities. In essence, you would have someone with a female brain."

It's not hard to understand why some people have been reluctant to acknowledge the differences. For generations, those differences have been used to define women as less competent, less intelligent, and less valuable. And that painful experience has at times discouraged not just research, but even honest conversation.

It seems to me there are two ways to approach this conflict.

You can deny the differences. If you claim they don't exist—or are mere cultural constructs—then it makes sense to create a world in which people, regardless of gender, are expected to think and act alike and are judged according to a uniform standard. But that solution ignores biological reality: There is no unisex brain, there is no unisex norm. There is only the male norm. And it undervalues the powerful, sex-specific strengths and talents of the female brain.

The alternative is to embrace the differences. According to Brizendine, women have "outstanding verbal agility, the ability to connect deeply in friendship, a nearly psychic capacity to read faces and tone of voice for emotions and states of mind, the ability to defuse conflict. All of this is hardwired into the brains of women." In the end, it may not be the lack of intrinsic aptitude that keeps more women from pursuing careers in fields like physics and engineering; rather, it may be that their different realities compel them to make different choices. It's time that those realities were recognized as every bit as relevant, important, and valuable as men's.

"Women have a biological imperative for insisting that a new social contract take them and their needs into account," Brizendine explains. "Our future and our children's future depends on it."

Part II

Why Women Should Rule the World

Chapter 4

If the Three Wise Men Had Been Women

"My second favorite household chore is ironing. My first being hitting my head on the top bunk until I faint."

—Erma Bombeck

In a lot of ways, my dad has lived his life in two parallel universes. His home life has been an all-girl affair. Growing up, he had four sisters and no brothers, and when he was eleven, his dad died. After college, he married my mom and promptly had three daughters. Even our dog was female. At the same time, his work world was a virtual boys' club. In high school, he worked at the family gas station, and in college, he spent summers as part of a road crew paving the Indiana toll road. For twelve years, he was a carrier pilot in the United States Navy, and then went to work for the defense contractor Lockheed, test-flying, building, and marketing airplanes. Boys by day, girls by night—and ne'er the twain shall meet. The fact that he had so little contact with women in the "real" world might help explain why he always seemed a little stumped by what might be considered "normal" female behavior. "What could you possibly have to talk about for three hours?" is still a question my dad is likely to ask after a particularly lengthy lunch involving any of the women in his life. And while he would never try to stop such a gabfest, neither does he understand its value; he just can't seem to shake the feeling that after a certain point, the time would be better spent washing the car.

I used to think he was being stubborn. But I now realize it's just the way he's made. When the news is bad—or good—he doesn't feel the urge to pick up the phone or put on the kettle and share the news (and a cup of tea) with a friend.

A few years ago, a couple of women scientists were in a lab at UCLA, when they got to discussing the different ways they and their male colleagues dealt with their demanding jobs. "There was this old joke that when the women who worked in the lab were stressed, they came in, cleaned the lab, had coffee, and bonded," said Dr. Laura Cousino Klein, one of the scientists. "When the men were stressed, they holed up somewhere on their own." Dr. Klein and fellow researcher Dr. Shelley Taylor wondered if there was more to their observation, so they decided to look at the data. They found that up to that point, the vast majority of stress research had been conducted on men. But as soon as they started teasing out the findings on women, "the two of us knew instantly we were on to something," Klein explained.

That something was a landmark study that showed that men and women manage stress in remarkably different ways. When men are stressed, they get in someone's face—or retreat into their proverbial caves. Women also experience this fight-or-flight response to stress. But as Drs. Taylor and Klein reviewed hundreds of studies and talked with scientists from a wide variety of fields, they found that women were more likely to respond to stress in their own way: by hanging out with their kids or talking things over with a friend or family member, a pattern of behavior which they called "tend and befriend."

Why? At least part of the reason appears to be biological. When they are stressed, both men and women produce a hormone called oxytocin, which has been shown to increase bonding and decrease blood pressure and anxiety in some animals. But estrogen—which is produced in much greater quantities in females—actually boosts the effects of oxytocin (also known as the "cuddling hormone"), while testosterone, the male hormone, blocks it. So when women are stressed, their brains start cranking out the oxytocin, which compels them to reach out to their friends and families, which helps them relax, which stimulates the production of even more oxytocin. Next thing you know, they're cleaning the lab, drinking coffee, comparing notes—and making scientific breakthroughs.

The simple fact is: Men and women often experience the world differently. Regardless of the reasons, which are doubtless complex (as discussed in chapter 3), the effects are real. They influence what we buy, what we read and what we watch, who we vote for, and how we spend our time. They shape our priorities and values. They make us who we are. And when we include and respect these different points of view—regardless of whether the lives that shaped them came wrapped in a pink or a blue blanket—we broaden the dialogue, expand the scope of inquiry, change the way we think. We make business more efficient. We make government more responsive. We get better art, better science, better schools. In short, everybody wins. But it starts with accepting the differences.

Sometimes the difference is in the details, as Elena Kagan, the first female dean of Harvard Law School, discovered a few years ago. While she was attending a conference, she stepped into the women's bathroom and found a basket neatly stacked with free tampons. "It's such a small thing, but it says a lot about whether a place cares about its women," she said. A few months later, every women's bathroom at Harvard Law had a supply of tampons.

COMMON SENSE AND THE BOTTOM LINE

Over the past couple of decades, women have made astonishing gains in Hollywood, and they're now well represented in nearly every job category: writer, producer, director, deal maker, studio chief. There are a lot of reasons for their success, including a culture that was created by outsiders and may have been more open to them. But they also brought a different perspective—and that meant they often wanted to make films about different subjects, that appealed to different audiences, that tapped new sources of revenue. "No one thought anyone would go see *The First Wives Club*," Sherry Lansing, the former chief of Paramount, told me, referring to the 1996 comedy starring Goldie Hawn, Diane Keaton, and Bette Midler about three divorced women who seek revenge on their husbands for marrying younger women. "We partnered every movie; that was the philosophy of Paramount. That was the only movie I couldn't get a partner on. No one would partner it. And it was a hit. People were shocked. And I said, 'Why are they shocked? I go to the movies. Why would they be shocked?'" And that started the "female empowerment" chick flick, which was enormously successful.

Lansing explained that filmmakers choose projects that "reflect what's going on in your gut and what you think. So I know why I wanted to make *Fatal Attraction*, because I read about Jean Harris"—the headmistress of an elite girls' school in Virginia, who in 1980 murdered her lover, diet doctor Herman Tarnower, in a jealous rage. "And I remember when some guy left me, and I couldn't get out of bed. And I was like, 'I have a college degree, what is wrong with me?' I'm dialing his number and hanging up the phone and driving by his house and I thought I was losing my mind. And so this story fascinated me. And I understood the Glenn Close character."

Who didn't occasionally dial someone's phone number and hang up, back in the days before caller ID? Glenn Close played the villain, a single working woman who went nuts when a married man with whom she'd had a very brief affair, played by Michael Douglas, refused to return her calls. The moral center of gravity, by contrast, was the wronged wife, a stay-at-home mom played by Anne Archer. Some feminists saw the setup as a screed against the evils of ambitious women, and the backlash was strong—and unexpected. Lansing said she was "devastated" by the reaction. "We thought it was a very profeminist thing because a guy would never sleep with a woman again and not call her back, which I thought was like the most unbearable thing he could ever do," she recalls.

Since roughly half of moviegoers are women, it makes sense that something approaching half of moviemakers would also be women. And while we're not quite there, the trend line is definitely positive. But what about other businesses?

Judith McHale, the former CEO of Discovery Communications, believes that a corporation should reflect the community it serves. "So if 50 percent of your viewing public is women, how else do you get the voice of that part of the community that you're serving?" she asks.

Women make the vast majority of consumer decisions, not just in the United States, but around the world. So does the management of most companies reflect the makeup of their consumer base? In a word, no. Which seems ridiculous on its face. Why wouldn't you want to include women in discussions about how to make and market products that women buy like diapers, dish soap—and computers, men's clothes, and cars? (Yes, women buy more computers, men's clothes, and cars than men.) It makes even less sense when you consider that studies show that companies with more women in senior management are more profitable. (See "The Smart Thing" in chapter 2.) This isn't about what's PC; it's about what works.

And yet. Look at Revlon, a company that sells virtually all of its products to women. Quick: Name a Revlon product made for men? Last I checked, there were no metrosexual eye shadows or lip liners at my local drug store. But at the end of 2007, all of the company's senior managers and all but three members of its board were men. Ditto Procter & Gamble, the consumer goods giant that sells 85 percent of its products to women. Also as of 2007, only nine of the top forty-seven corporate officers were women—not reflective of the market, but more than there were just a few years ago.

Come on, people! It's not that women are so much smarter (though sometimes they are). It's just that they provide an enormous and, too often, untapped source of talent and bring a different perspective—which by definition creates opportunity.

Tom Peters, the management guru who shook up the business world in the 1980s with his mega-bestseller *In Search of Excellence*, believes women are critical to future growth and success of business worldwide. To illustrate his point, he told me a story about a speech he gave at a gem—as in jewelry—conference a few years ago. "Most of the jewelers are men. Obviously most of the eventual receivers thereof are women. And so these guys are telling me that, 'Well, men still buy most of the jewelry for women.'" But since women buy the vast majority of consumer goods, he couldn't understand why they weren't buying more jewelry. Finally, one of the most powerful women in the business explained things. "She says, 'Men buy most of the jewelry for women because men buy most of the jewelry for the stores, and men buy jewelry that other men would give to women.' She said, 'I go to shows and I buy jewelry that women would want to buy for themselves.' I thought, 'Oh, praise God,' you know?"

Women's different perspectives don't just affect the bottom line, however; they also move the culture. Debra Lee, the CEO of Black Entertainment Networks, told me that her own life experience had a significant effect on the way she led the company—once she let it. Not long after Bob Johnson, BET's founder and CEO, promoted Lee to president and COO, the head of human resources called her to ask if the company would participate in "Take Our Daughters to Work Day," something they'd never done.

"And so my first inclination was to say, 'No.' I was like, 'Well, we've got a lot of work to do. We can't have kids roaming around the hallways,'" she explained. "But then I thought about it overnight. And I realized that what I was doing was answering the question the way I thought Bob would answer it. I thought, 'I'm COO now. And I've got to run this company. And we've got to get profits, blah, blah, blah. So I've got to keep a handle on things.' And then I thought, 'Well, if I do that, why am I here? If I don't realize there's something different about me, and the way I manage, and what I think is important, and what I'm going to bring to this company, I shouldn't be in this position.' So I came to work the next day, called the head of HR back, and I said, 'Look, we're going to do it.' I said, 'They can't stay past two o'clock.' I mean I had all these requirements," she said with a laugh.

Ten years later, Take Our Daughters to Work Day has become a thriving tradition at BET. "It's a huge project for us every year," she says. "We get eighty to ninety kids, and it's boys and girls. And they come in, and we have a full day of activities. We show them how to broadcast. They sit behind a news desk, and we tape them. And I go over and speak to them. It's really an established program. And the kids love it. And the parents love it.

"And that's just one little example where I said, 'Okay, I've got to step out of this *I've got to manage like a man*.' This is important to me because it's for girls. And there's a way to make it work—and still achieve the company's goals. And I think that's what women have to do."

DIFFERENT PERSPECTIVES, DIFFERENT PRIORITIES

As women have played an increasingly important role in politics, there is no question that they've brought a different perspective, focusing attention on a broader set of issues and building alliances with other women. That's not to say that women in elective office focus only, or even mostly, on so-called women's issues. But research confirms that both Republican and Democratic women are more likely than their male counterparts to initiate and fight for bills that champion social justice, protect the environment, advocate for families, and promote nonviolent conflict resolution. They also focus on issues like transportation, agriculture, and arms control, just like men. But women, as Geraldine Ferraro once said, "raise issues that others overlook, pass bills that others oppose, invest in projects that others dismiss, and seek to end abuses that others ignore." Amen.

Kay Bailey Hutchison was the first Republican woman elected to the Texas House in 1972. "There were four Democratic women and me, for a total of five who were elected that year. And we did get together to do several things that were definitely a result of our experience as women. We changed the laws regarding rape victims in Texas and became really the leader in the nation on fair treatment for rape victims. And we did that as a coalition. We did equal credit rights for women. We did historical preservation. I also did transportation . . . That was not my experience as a woman, but I definitely worked with the Democratic women. And it was a great coalition that we had because the Republicans knew that if I was on it, it was okay. And the Democratic women had the credibility with the Democrats. So when we went together, we just mowed over them."

Hutchison believes the goal of representative government is to bring together as many backgrounds, points of view, and experiences as possible "to make a better result. And it's just that, historically, the women's experience was not at the table." But that's changing, she told me during an interview in her stately Senate office, and the effect on legislation is undeniable. "The homemaker IRA was my deal because of an experience I had as a single woman starting an IRA when I was working," she explained. But after she got married, moved from Houston to Dallas and temporarily stopped working, she found she could no longer contribute to it. "And I said, 'Well, what is this?' And so you have that experience and you bring that experience to the table. So our bill did correct that so women who work inside the home now have the same retirement opportunities on IRAs as women who work outside the home."

Kathleen Sebelius, the Democratic governor of Kansas, told me she's a "huge believer" that more women in elective office would produce better decisions. "People bring their own life experiences, and women's life experiences are different than men's—not better, not worse, different. And 51, 52 percent of the population is women. And so having people at the table who make decisions based on their life experiences, their lens—whether it's as a mother, a daughter, a spouse, somebody who's in the workplace—I think we get better policies, a better dynamic."

Sometimes women just think to ask different questions. After her children were grown, Lindy Boggs—a congresswoman from Louisiana who had won her husband's seat after he was killed in a plane crash—decided to sell her house and buy a condominium closer to Capitol Hill. When she went to the bank to get a loan, she brought a financial statement showing that her congressional salary was more than adequate to cover the payments. Nonetheless, the loan officer asked her for detailed financial statements and insurance policies, saying they were required by federal law. "Until then, I had not personally experienced the situation women often faced when they applied for a bank mortgage and did not have a working husband behind them," she wrote in her autobiography, *Washington Through a Purple Veil.* But Boggs knew better.

"My dear," she replied, "I am the author of the law that forbids this type of requirement for female persons and the elderly. You are not complying with the federal regulation, you are in defiance of it." Game, set, match.

Years later, her daughter, ABC News and NPR correspondent Cokie Roberts, went to the bank to refinance the home she had earlier purchased from her mother. According to Boggs, she was told by the banker, "Oh, this form is nothing, it's just a boilerplate saying we haven't discriminated against you because of sex, race, etc." Cokie interrupted: "Nothing! It's not nothing! My mother wrote that law!"

Not long after I left the White House, the surgeon general was forced to resign in a sex education controversy, and President Clinton nominated Dr. Henry Foster, a Tennessee obstetrician with a sterling reputation, to take her place. The White House vetting team had reviewed Foster's records and questioned him extensively, and he had passed with flying colors. But just days after the president announced his nomination, it became clear that the president's staff either hadn't asked the right questions—or hadn't shared all the relevant information. Dr. Foster, it turned out, had performed a small number of abortions across his more than thirty years of practice. And while it was a tiny part of his admirable and public-spirited career, it exploded into a controversy that ultimately sunk his nomination. The U.S. Senate refused to confirm him. How could the White House have been caught so unprepared? After all, abortion had been a profoundly divisive issue since the Supreme Court had legalized it in 1973. One possible answer: There were no women in the room when the questions were asked, the answers considered, and the strategy of how—or even whether—to proceed discussed. Maybe two or three or ten women wouldn't have changed things. But given the different experience women bring to issues involving obstetricians, isn't it more likely that they would have?

Women's particular experiences continue to shape not just their points of view but their actions, in the United States and around the world. In 2004, Wangari Maathai became the first African woman to be awarded the Nobel Prize for Peace when she was recognized for her contributions to sustainable development. Through the Greenbelt Movement, the organization she founded some thirty years ago, Maathai has helped plant more than 30 million trees across Africa, providing firewood, clean drinking water, balanced diets, shelter, and income for rural families.

While still a graduate student, Maathai attended a forum of the National Council of Kenyan Women. There, she explained, "I listened to the women from the countryside, you know, who were leaders, and the women said that their biggest problem was that they did not have energy, which is collecting firewood. They did not have clean drinking water. They did not have adequate food. And they did not have an income. And that attracted me because these women were coming from the same countryside I was brought up in. Just about ten years earlier, I was a child, a young girl, running around that same countryside. And at that time, we had plenty of firewood, we had plenty of clean drinking water, we had plenty of food. We didn't have a couch, as such. No income for a couch. But we did not think of ourselves as poor. So I became interested in what was happening to the countryside."

She had been there: collecting the firewood, fetching the water, tending the crops. And she understood how difficult life would be without these basic necessities. And so the movement was born. At first, it was all women. "And for a very long time, men were not involved. But they let their wives participate. It took much longer for the men to respond and start participating. And they participated when they saw economic returns . . .

Because when the women planted the trees and they survived, we gave them a small financial compensation. So they did see financial compensation. But in the beginning, the men didn't see it, because it was a very small financial compensation for a lot of work. Now that's so common with us that we will take it!" she says, with a knowing laugh.

As women slowly gain power, their values and priorities are reshaping the agenda. A multitude of studies show that when women control the family funds, they generally spend more on health, nutrition, and education—and less on alcohol and cigarettes. The effects extend beyond the family. In one study of local councils (*panchyats*) in India, researchers found that when women are in charge, they make different choices than men, investing in projects that directly affect their particular needs, like clean drinking water and better roads.

That's not to say that women's priorities are better than men's. Rather, when women are empowered, when they can speak from the experience of their own lives, they often address different, previously neglected issues. And families and whole communities benefit.

The Mommy Brain

One of the profound and life-altering experiences that most—though not all—women share is having children. Obviously, men have children, too. And it's profound and often life-altering for them as well. But there's no denying the biological realities of motherhood. Women spend months in various stages of adjustment: morning sickness, heartburn, leg cramps, multiple sets of (unflattering) maternity clothes, more heartburn, sleeping with a watermelon grafted onto your abdomen, as well as joy, wonder, anxiety, fetal hiccups—and a cleaning frenzy that attacks every dust bunny like it was a member of the Taliban. And that's all before the baby is even born. Once the baby comes, life really starts to change, as 3:00 a.m. feedings give way to nailing the toilet seat shut to keep the kid from trying to flush the latest three-pound issue of *Vanity Fair*. Each stage of motherhood brings new joys, frustrations, fears, and conflicts. And while every woman experiences those in different ways and on a sliding scale of intensity, children become—for virtually every woman—the pillar around which the rest of her life is built. That's not to say that plenty of women, even most women, won't continue to work, at some point, either full time or part time. It's just that before they walk out the door every morning, they'll have to make sure that their kids are taken care of. And virtually every decision they make will start with the question, "How will it affect my children?" It's that simple—and that all-encompassing.

To be sure, men help. Some more than others. Every family needs to find a division of labor that works for it. But let's be honest: The majority of women are—and will continue to be—the primary caretakers for their children. Studies show it. And experience confirms it.

Before I had children, I wasn't very interested in them. I knew I wanted children—and somehow, I always believed I'd have a couple. But I wasn't in a hurry to get on that track (I was thirty-eight by the time my first child was born). Now I am the mother of two, and the experience has utterly changed me. Not just my life (so long, spontaneity!). But me. New research on women's brains suggest that motherhood literally alters a woman's brain "structurally, functionally, and in many ways, irreversibly." In her book *The Mommy Brain: How Motherhood Makes Us Smarter*,

Katherine Ellison argues that the constant physical and emotional demands of raising children—and the flood of hormones that goes with it—actually improve women's senses such as smell, taste, and touch; sharpen their social skills; and make them more efficient, resilient, and motivated.

That's certainly been my experience. My husband has an infinitely more fine-tuned sense of smell than I do: He can sniff out a piece of ripe cheese in the fridge the minute he walks through the front door. (So much so that I often accuse him of having olfactory hallucinations.) But my ability to hear—scratch that, *feel*—what my kids are up to far exceeds his. The second one of them sets foot out of bed in the middle of the night, I know it. And while I wasn't exactly a social retard before I had kids, my ability and my desire to connect with people, to pick up their verbal and nonverbal cues and to really tune into what they're *feeling* is profoundly better than it was back in those footloose and child-free days. I'm also dramatically more patient. And I've become a relentless multitasker; I *never* walk up or down a flight of stairs in my house without a load of laundry, mail, toys, groceries, or garbage in my arms.

When my son, Stephen, was not quite two, he introduced me to a whole new dimension of crisis management. I was checking the chemicals in our small backyard pool—a task I usually avoided when he was around for obvious reasons—and he was playing a couple of feet away. The pool needed a little chlorine, so I opened the plastic container, then set it down a few inches from my hand while I double-checked the level. When I turned back—literally seconds later—he was holding the two-pound container like a cup and pouring the powder into his mouth. His eyes bulged as it reached his throat, and suddenly, everything shifted to slow motion. I felt the panic rise. My son started to cry, as I grabbed him and ran for the house, screaming I think. My husband just happened to be home, so I ran upstairs and yelled for him to help. I found the Poison Control number pasted inside a kitchen cabinet and tossed it to my husband, while I sat Stephen in the kitchen sink and started spraying out his mouth, trying to keep the chemicals away from his eyes. He started throwing up. I kept rinsing. I peeled off his orange polo shirt, which already had a huge bleach stain on the front. *If it's taking the color out of his shirt, what's it doing to the insides of his little body,* I wondered, not daring to really consider the possibilities. I kept rinsing. By now my husband had the poison control center on the phone, and the incredibly calm, knowledgeable operator talked us through the next steps. "She says to see if he'll eat or drink anything," my husband said. Somehow, Stephen had stopped crying and throwing up, so I gave him a cup of water. He took a few sips, while I dried him with a dish towel. It didn't seem to hurt when he swallowed, so I gave him a Popsicle. He took a few licks, and again, he didn't seem to be in any pain. That's good, the woman on the phone assured us. It may mean he hasn't burned his esophagus. I gave him a cracker; he ate it. Still no sign of pain.

The immediate crisis had passed, and Stephen was fine. In fact, he kept a date to go to the zoo. We were incredibly lucky. But the fact that he was able to swallow chlorine on my watch still haunts me. And I found I wasn't as cool in the crisis as I thought I'd be. More importantly, I realized how parents, particularly mothers, are forced to make critical, sometimes life-and-death decisions immediately. No net, no second chance. And that kind of experience can change the way women see themselves in important and permanent ways.

In so many relationships, women are expected to defer to other people in their lives: their parents, their husbands, their bosses, their neighbors, the checker at the local grocery store. But when children are involved, they're the boss, the go-to person, the (as George W. Bush would say) "decider." The psychologist Daniel Stern has specialized in the study of mothers. "The expression 'the buck stops here' takes on new meaning," he writes in his book *The Birth of a Mother.* "You will have to make split-second decisions even when you don't really know what to do and haven't been there before. It's akin to being a CEO, a policeman on duty, or a physician on call. All eyes turn to the person in authority and expect that person to know what to do."

Give a Woman an Inch

Naturally, not every woman is a mother, and not every mother experiences her child-rearing years in same warm glow of hugging and learning. Real life is often more like the Simpsons than the Waltons. The point is: Motherhood was for too long something that women felt they had to minimize or even hide if they wanted to be taken seriously in public life. Don't talk about your kids, they were told; don't do anything to remind people that you're not a man. Well, guess what? Even when women denied the life-altering realities of being moms, they couldn't change them. They still weren't men. Now women are rightfully claiming motherhood as a formative experience—an education, like law school or business school—that teaches important skills and prepares them to lead.

Being a mother doesn't always make women more sympathetic to the plight of other women or other mothers. We've all heard the horror stories about the cold-hearted female bosses who are anything but patient and empathetic. But I've heard vastly more stories about how women's own experience has made them sensitive to the challenges other women, particularly mothers, face.

Shirley Tilghman had been a single mom with two children and a demanding career as a professor and scientist. And when she became president of Princeton, she was determined to do what she could to make the university more family-friendly. Among other things, she realized that many young women saw the timing of tenure as an obstacle to their success. At the very time many had young children at home, they were expected to spend more time in the lab and in the classroom in order to make tenure within the allotted five years. So Tilghman changed the rules, allowing professors to request an extra year. But pretty soon she realized that a lot of women were reluctant to request the extension, fearing that it would make them seem less committed, less serious. So Tilghman made it mandatory. Professors can request that they be reviewed for tenure a year early, but the expectation is that everyone will take six, rather than five years. "That was very symbolic," she told me. But she also tried to take steps that were practical.

"We now have a program called Back-Up Care. You can call this service at seven-thirty in the morning and say, 'Can you have someone here at eight-thirty? My kid has a temperature, and I have a lecture at nine.' For four dollars an hour. We subsidize it," she tells me. "You can not imagine the kind of kudos we are getting." But the program isn't just aimed at parents with child-care issues. "My other example is the Maytag refrigerator repair man," she explains. "They say they'll be there between nine and twelve, and you think: *This is not helpful.* So you call this service, and they will sit at your house and wait for the refrigerator repair man. For four dollars an hour."

A couple years ago, my friend Colleen sent me a hand towel that said:

If the three wise men had been women, they would have asked directions, arrived on time, helped deliver the baby, cleaned the stable, made a casserole, brought practical gifts, and there would be Peace On Earth.

Of course! But they weren't women, so the wise men got there late—and brought gold, frankincense, and myrrh. Can you imagine what Mary—who's just delivered a baby on a bale of hay without an epidural—must have thought about that?

The truth is: Women are just more practical than men. As Geraldine Ferraro says, "Give a woman an inch, and she'll park a car in it." Maybe it's biological. Women use multiple parts of their brains to accomplish certain tasks, like talking or solving problems, while the same activities in men's brains tends to be more focused in one particular area. Maybe that helps explain why a man who is focused on reprogramming the DVD player might not notice the house is on fire, but a woman can take a conference call, feed the baby, unload the dishwasher, marinate a flank steak, and rebalance her investment portfolio all at the same time.

Senator Dianne Feinstein explains this practical quality in women another way. "Women are accustomed, I guess, to cleaning up after men. And in a sense, it's one of the reasons I think women make very good mayors and governors. Because they are very management oriented. They know things have to be managed, that you have to stay the course. You have to see that it's done every day, that you lay out your checklist, and you go through it. Women are accustomed to that from the managing of the house."

Because of their different lives, women often bring different strengths, different priorities, and different values to public life.

"You really see that in a part-time legislature, where people are coming in and out of their lives," Kathleen Sebelius explains. "Well, the lives that the women are coming in and out of are very different from the lives the men are coming in and out of."

That's not to say that there is a one-size-fits-all "woman's life." A lot of working women have more in common with the male executive down the hall than the full-time mom across the street. Still, the differences are there.

For too long, experience that was uniquely, or even mostly, female, was devalued; unless men shared it, it just didn't count. It's time to move beyond a place where men decide what women want and what they will buy. Where policy is made without consulting the people most affected—or those who are left out. Where football—but not motherhood—is considered a suitable training ground for leadership. The more women own their experience—whatever it is—the more likely they'll be to succeed, in their own ways, on their own terms.

Chapter 5

The Nature of Violence

"A nation is not conquered until the hearts of its women are on the ground. Then it's finished, no matter how brave its warriors or how strong their weapons."

—Cheyenne proverb

As I sat in the darkened theater, I felt a rising sense of dread. *Was that me?* I wondered. My husband, Todd, and I were watching *Hotel Rwanda*, a powerful and disturbing film about the 1994 genocide in central Africa. In one scene, a radio blares an excerpt from an American government news briefing.

"Based on the evidence we have seen from observations on the ground, we have every reason to believe that acts of genocide have occurred in Rwanda," says a female briefer.

"What's the difference between 'acts of genocide' and 'genocide'?" a reporter interrupts.

"Well, I think . . . as you know, there's a legal definition of this . . . clearly not all of the killings that have taken place in Rwanda are killings to which you might apply that label," she continues. "But as to the distinctions between the words, we're trying to call what we have seen so far as best as we can; and based, again, on the evidence, we have every reason to believe that acts of genocide have occurred."

"How many acts of genocide does it take to make an actual genocide?" the reporter shoots back, clearly disgusted.

The questions are tough, but the answers sound both ridiculous—and eerily familiar. *That's not my voice,* I tell myself. *But are those my words? Was the exchange based on something I said while I was White House press secretary?* I tried to concentrate on the movie, but my thoughts kept circling back to the tape. *Let it not be me. Let it not be me.* As soon as we got home, my husband searched the Internet and found that it was an actual audiotape from a State Department—not a White House—briefing. I was both relieved—and certain that I had said some of the same things.

It had all started on the evening of April 6, 1994, when Hutu extremists shot down a plane carrying the president of Rwanda, also a Hutu, killing everyone on board. The saboteurs had opposed a peace plan the president was implementing to end the civil war, and by nightfall, they and their allies among the Rwandan armed forces and Hutu militias had set up roadblocks and were going house to house killing Tutsis and moderate Hutus who supported the agreement. Thousands died that first night.

In the weeks that followed—as the killing raged—the United States and United Nations repeatedly condemned the violence, but stopped short of sending troops. The death toll soared into the tens, and then hundreds of thousands; most of the dead were Tutsis, killed simply because they came from the wrong tribe.

Still, the situation was not a hot topic in the White House briefing room. Continuing violence in Bosnia; crime and health-care bills working their way through Congress; and a vacancy on the Supreme Court got top billing. As the president's press secretary, I didn't get many questions about what was happening in Rwanda.

But when I did, it was agonizing. Much of the discussion centered on the definition of "genocide." And the debate wasn't merely semantic. If what was happening in Rwanda was, in fact, genocide, then the UN by its mandate and the United States by its morals would be required to intervene. Instead, American government officials and their spokesmen—like the woman in the movie—could say only that while "acts of genocide" had clearly been committed, it could not yet be characterized more broadly as genocide. It was the kind of legal hairsplitting that can make life on the podium miserable.

Finally, in mid-July, the Tutsi-led Rwandan Patriotic Front captured Kigali, the capital, and what was left of the Hutu-led government fled to Zaire, followed by a flood of refugees. Disease and more killings in the camps followed, but the genocide was over. More than 800,000 people had died in just 100 days.

Today the Parliament in Kigali is one of the few buildings with visible scars of the fighting that ended more than a dozen years ago. Outside, the bullet holes bear witness to the killing that shocked, but didn't mobilize, the world. But inside, there are signs of change—and of healing. Nearly half the seats in the Chamber of Deputies are held by women, a reality unimaginable before the war. In fact, women hold thirty-nine out of eighty seats in the lower house—49 percent—more than any other country, anywhere on the planet. They also hold twenty of sixty seats in the upper house. Together women account for more than

42 percent of the total; worldwide, only Sweden and Norway have more. (By contrast, the U.S. Congress is roughly 16 percent women, an anemic eighty-first in the world.)

Rwanda's continuing recovery and women's increasing power are inexorably linked; the violence made a new social structure both necessary and indispensable to any hope of reconciliation and a more peaceful future. The country was left with a population that was overwhelmingly female. "Before the genocide, women always figured their husbands would take care of them," Aurea Kayiganwa, the coordinator of organization of war widows, says. "But the genocide changed all that. It forced women to get active, to take care of themselves. So many of the men were gone."

So the women did what they had to do: They buried the dead, healed the wounds, and began to rebuild civil society. "We were the wives left as widows, the mothers whose children died. We are the owners of the post-war issues," says Aloisea Inyumba, a Rwandan governor, former cabinet minister, and former head of the National Unity and Reconciliation Commission. One of the first issues they faced was the hundreds of thousands of children orphaned by the bloodshed. Inyumba, a leader in the effort to find them homes, worked to convince families to take children, regardless of tribe. "It was like asking Jews to take in Germans after the Holocaust," Swanee Hunt, a peace activist, philanthropist, and former U.S. ambassador to Austria, told me later. And they did.

Still, there were deep and painful issues dividing the women. "Widows of the genocide had to live, as best as they could, alongside women whose husbands were in prison accused of genocide crimes," says Esperance Mwiza, who served on the Women's Council and in parliament. "Then there were women like me, back from exile, returning to a country they didn't know, but who were very enthusiastic about having a country at last, and who wanted to buckle down to rebuilding it. It was not easy to reconcile the realities of these women who were separated from each other by history."

But reconcile they did, and in the process, women became symbols of moderation. They weren't blameless in the rash of violence: some killed, tortured, informed, collaborated, and otherwise participated. But for the most part, women were not "the planners or perpetrators." Only 2.3 percent of the more than 100,000 people jailed for their roles were female. And because women were substantially less involved, they were able to take on critical roles in the reconstruction. "There is a widespread perception in Rwanda that women are better at reconciliation and forgiveness," says Elizabeth Powley, who authored a study on Rwandan women for Women Waging Peace.

As the social structure began to change, so too did the laws. Old statutes prohibiting women from owning or inheriting property were rewritten, giving them more control over their own and their families' financial well-being. "Traditionally, a woman is not a breadwinner; now she has had to become one," says Inyumba, who was Minister of Gender during the law's drafting process.

In 2003, women banded together to help shape a new constitution—one that greatly expanded their rights. Among other things, 30 percent of the seats in the two houses of parliament were set aside for women. Such quotas would be impossible in the United States, where policies aimed at increasing the representation of women or minorities in public institutions are politically perilous and legally suspect. But when Rwandan voters went to the polls in October of 2003—for the first time since the genocide—they chose more women than the law required, surprising a lot of male politicians and leading to near parity in parliament. "Some men even complained that women were taking some of the 'men' seats," said Donnah Kamashazi, a representative in Rwanda for the United Nations Development Fund for Women.

Women's expanding power isn't limited to parliament. Women are also serving on the Supreme Court, in the president's cabinet, in the senior ranks of the police force, and on the influential National Unity and Reconciliation Commission. Their influence is nowhere near that of their male peers; Rwanda remains a deeply patriarchal country. Nor have the myriad problems that slow down progress been solved; far from it. But women are making progress—and in the process, they are healing their country.

"After the genocide, women rolled up their sleeves and began making society work again," said Paul Kagame, president of Rwanda.

Building Bridges

Women are becoming key players in peace processes, not just in Rwanda but around the world. "For generations, women have served as peace educators, both in their families and in their societies. They have proved instrumental in building bridges rather than walls," says Kofi Annan, former secretary general of the United Nations.

A variety of studies have found that because women are generally less violent than men, they are often better at creating and keeping the peace in post-conflict societies. A recent World Economics Forum study looked at Cameroon, Bolivia, and Malaysia and concluded that when women had more control over spending, they spent less on the military. In addition, a series of studies by Harvard psychologist Rose McDermott found that the more money a country spends on its military, the more likely it is to go to war. Basic geometry teaches: If A equals B and B equals C, then A equals C. In other words, more women mean fewer wars. Maybe it's not that simple. Then again, maybe it is.

As I look around the world, a number of things seem clear. First, women often bring a different point of view and different priorities to questions surrounding war and peace. And that perspective can be crucial to ratcheting down the violence, creating opportunities for reconciliation, and beginning the process of rebuilding.

"After the war in Bosnia, I asked the prime minister, Haris Silajdzic, 'If half the people around the table at the very beginning had been women, would there have been a war?'" Swanee Hunt tells me. "And he said, 'No. Women think long and hard before they send their children out to kill other people's children.'"

Recently, Sally Field tried to make essentially the same point. In accepting an Emmy Award for her role in the television series *Brothers & Sisters*, she said: "Let's face it: If mothers ruled the world, there would be no goddamn wars in the first place." The millions of people watching at home never heard the end of her sentence. Fox, which carried the program live, argued that it cut the audio and video feeds because Field used a profanity; the Federal Communications Commission has cracked down on offensive language, and broadcasters are slapped with heavy fines when prohibited words make their way onto the air. But a lot of people believed that Fox objected to the actor's antiwar tone.

Politics and profanity aside, Field has a point: Mothers have a lot to lose when their children go to war. That's not to say that they're all pacifists. From the Peloponnesian Wars of ancient Greece to the jihad of today's Islamic fundamentalists, many a mother has willingly sent her sons (and occasionally her daughters) into battle, ready to lose them to the higher cause. My own father, a Navy pilot, spent years in harm's way, including two long deployments in Vietnam. Had he been killed, I'm sure his mother (as well as his wife and daughters) would have been both devastated—and proud. And I often wonder how I might feel if my own son decides to follow in the footsteps of the grandfather he idolizes, choosing to guard the ramparts and give his life if necessary. Still, I think mothers have a different standard for measuring the costs of war—and the price of peace.

Securing the Peace, a report by the United Nations Development Fund for Women, called it "essential" that women participate in every phase of the peace process, from talks, to implementation, to monitoring. "When approaching the difficult task of ending war, the stakes are too high to neglect the resources that women have to offer," the report says. A quick survey of countries recovering from conflict supports that conclusion.

A few years ago, I met with a group of women from Mostar, an ethnically divided city in Bosnia and Herzegovina. They had recently formed a Women's Citizen Initiative to further the work of bringing the city together after the war, and they were in Washington to accept the Madeleine Albright Grant from the National Democratic Institute for International Affairs. When the women began their work, Croats and Muslim Bosnians were so divided that the city's high school had two floors, two curricula—and two separate entrances. I asked one of the leaders, Amira Spago, who had lost her husband in the conflict and was left to raise her two young daughters alone, why it took women to reach across the ethnic divide. "Women are more practical," she told me. "We're really ready to change." It's not that men don't cooperate, she said. "They do business together. They form partnerships in business. During the war, they sold weapons to each other. They will do everything for their personal interest."

In Northern Ireland, women's groups spent a decade building trust between Protestants and Catholics, before they were finally allowed to participate in the peace process itself. At one point, male negotiators walked out of the talks, leaving a handful of women—including Monica McWilliams and other members of the Northern Ireland Women's Coalition—at the table. They kept the conversation going—and the Good Friday Agreement was completed, ending a conflict that had lasted decades and claimed thousands of lives. "Men are stubborn," says McWilliams. "Women are more comfortable seeking compromise. They see it as a strength, not a weakness."

During South Africa's post-apartheid transition, black and white women were instrumental in planning, developing, and implementing the successful Truth and Reconciliation Commission. Its mission was to establish a "truthful" record of the apartheid era, giving people from all sides the opportunity to tell their stories, forgive their former enemies—and to be forgiven. The women insisted that TRC be inclusive and transparent, setting it apart from previous truth commissions and ensuring the greater participation of women. When all was said and done, more than 40 percent of the commissioners and three-quarters of the staff were women, and women provided more than 56 percent of the testimonies. Women witnesses often addressed the suffering of others: sons, husbands, neighbors, friends; it was less "me" and more "us," helping create a sense of shared purpose. The commission also created separate structures for women, making it easier for them to talk openly about issues like crime and sexual assault, and to help assure that the record was complete. The South African TRC has become a model for societies facing similar transitions.

OTHER VOICES, OTHER VIEWS

Despite women's increasingly important roles in resolving conflict and sustaining peace, their voices are too often dismissed, a reality that is as frustrating as it is predictable.

Pat Mitchell, one of the first women correspondents for the *Today* show, was hired first and foremost because she was a seasoned journalist. But she was also hired—or so she thought—to bring a different perspective to the broadcast. In the late 1980s, she bumped up against the limits of that open-minded ideal.

"I convinced Thomas Friedman [then executive producer of the *Today* show and no relation to the *New York Times* columnist] that I wanted to do a piece on how women leaders are different than men. And where in conflict zones, in particular, women could be observed making decisions that were different than men. And in that way, could they get us closer to peace? He thought it was a good idea. He said, 'Where do you want to go first?' I said, 'I want to go to Israel. And I'll look at the Arab and Israeli women.' I got to Israel, and I got a call from the president of NBC News, who said, 'No, we don't think this is an important story. You need to come back home.' And I said 'Why not?' He said, 'Because these women aren't elected. They're just average citizens.' Anyway, I quit. I did it on my own, and that was the beginning of my production company."

Mitchell began by talking with Israeli and Arab women to see how they approached the intractable conflict in the region. Regardless of which side they were on, all had witnessed an unbearable amount of suffering, she said. "And those women had made a decision that as mothers, and wives, and community leaders, they were going to do something different."

At the time, it was illegal for Israelis to meet with representatives of the Palestinian Liberation Organization in Israel. So rather than break the law, the women found a place where such meetings were legal: the European Union in Brussels. They planned a conference called "Give Peace A Chance: Women Speak Out," and invited women from around the region: from Israel and the Palestinian territories; from Europe; and from Turkey, Jordan, and Egypt. A thousand women came.

"The very first night, Shulamit Aloni from the Knesset stood up," Mitchell told me. "And she said, 'Okay, we've only got three days here. And we have to get to a peace agreement. So let's make some rules. Nobody gives long speeches. Nobody stands up and does their list of grievances. I killed you. You killed me. You hurt my family. I hurt—none of that.' So there was no political stuff going on. The hard work was in finding the language that worked. At the end of the three days, seven paragraphs were drafted"—the Brussels Declaration—"that would end the territorial conflict, so to speak, in the Middle East, as it existed then."

But when the women went home, their solution was ignored—by the Knesset, by the Palestinian Authority, by the powers that be. And nearly two decades later, little has changed. Nonetheless, the Brussels conference became the foundation of relationships that have grown stronger over time. And the region's women are more convinced than ever that they are an indispensable part of the peace process.

"In matters of peace, the voice of women is clearer and brighter than that of men," says Aloni, former chair of the Meretz Party in Israel and minister of education and culture under former prime minister Yitzhak Rabin. "Men enjoy their manliness, they receive medals and trumpet victory; but women, after the battle, remain with the ashes, the mourning, the widowhood and the orphans. Without medals, they have to rebuild the family, the home, the community."

Among other things, the Brussels conference led to the creation of the Jerusalem Link, an organization created in 1993 to further the active involvement of Palestinian and Israeli women in achieving peace and social justice. "During my meetings with Palestinian women I noticed a clear distinction in the communication methods of the two genders," says Terry Greenblatt, an Israeli peace activist and early leader of the Jerusalem Link. "Men exhibit a tendency to see the world in black and white, 'war' or 'peace.' I know of at least 7,000 other options in the middle.

"For men, negotiation is a synonym to playing cards. They would assemble in a room with a long table, sit one against the other and try to conceal their cards as much as they can," Greenblatt continues. "They are inclined to treat the man in front of them as an opponent, not as a partner. Women, on the other hand, would assemble and sit at the same side of the table. We put the strife and pains in front of us, look at them courageously and come up with a win-win formula. The Palestinian woman with whom I converse would be my neighbor eventually. I have no intention or any interest in playing infantile games with her. Making sure she leaves the room with a good feeling is in my utmost interest."

Amneh Badran, Greenblatt's Palestinian counterpart, is convinced that previous agreements failed because the two sides were left to interpret unclear resolutions in different ways. "It is well known that women usually delve into the little details and that they would never be satisfied with half-baked situations," she says. "Unfortunately, that kind of attitude was missing in Oslo [the site of secret negotiations that produced the 1993 "Declaration of Principles" signed by PLO chairman Yasser Arafat and

Israeli prime minister Yitzhak Rabin at a White House ceremony later that year]. If they had only left the job for women to conduct those critical peace discussions, we would have completed the Oslo agreements with much better defined solutions."

Both Greenblatt and Badran say they are sometimes criticized for working with perceived enemies. "Palestinians ask me, 'Do you think there are peaceful people left on the other side? Do you really believe that a few women can make a change in this chaotic situation?'" Badran says. "Of course, if a man was acting in my position, he would probably have been heralded as a 'man of vision.'"

ECONOMIC EMPOWERMENT AND PEACE

If women are to participate more fully in peace processes around the world, they must be empowered more broadly in their own societies—economically, politically, and socially. And that, in turn, will produce a broad range of additional benefits.

According to Amartya Sen, the Indian Nobel laureate economist, recent efforts to empower women have focused on "well-being," improving their quality of life. But increasingly, he argues, they should focus on "agency"—that is, giving women concrete tools such as education, economic power, and property rights—that will not only continue to make their lives better, but also lead to broader societal change. In *Development as Freedom*, Sen writes that "working outside the home and earning an independent income tend to have a clear impact on enhancing the social standing for a woman in the household and in the society. Her contribution to the prosperity of the family is then more visible, and she also has more voice, because of being less dependent on others."

Educating women also produces large and measurable benefits. First, it substantially lowers child mortality rates, particularly for girls, in ways that educating men or reducing poverty do not. In addition, educated women tend to have fewer children; provide better health, nutrition, and education to their families; and earn more income than women with little or no schooling. When Larry Summers was chief economist at the World Bank, he argued that educating girls probably produced better returns than any other investment in the developing world.

According to Isobel Coleman, a senior fellow at the Council on Foreign Relations, "increases in household income benefit a family more if the mother rather than the father controls the cash." Studies from countries as varied as Bangladesh, Brazil, Canada, Ethiopia, and the United Kingdom suggest that "women devote more of the household budget to education, health, and nutrition—and less to alcohol and cigarettes." In fact, when women's incomes go up, child survival rates improve an astonishing *twenty times* more than if a man's income increases by a similar amount. Yes, you read that correctly: *twenty times.* And children's weight measures improve eightfold. Similarly, when women borrow money, they spend more on school enrollment, child nutrition, and health care than men do.

Thirty years ago, Muhammad Yunus founded the Grameen Bank—which means "Bank of the Villages" in Bangla—to serve the rural poor in Bangladesh. Yunus believed that giving small loans to people who could never qualify for traditional bank credit would fuel local development and combat poverty. He was right. And the movement he began—widely known as microfinancing—has improved the lives of millions. From the outset, Yunus deliberately targeted women: They tended to be poorer; they had even less access to credit; and they paid back their loans more reliably than men. Worldwide, there are now some 70 million microborrowers—and 80 percent of them are women. Studies show that women with even small amounts of capital get more involved in family decision making, are more legally and politically aware, and participate more in their communities than other women. They also suffer less domestic violence.

As women gain power, they help shape broader decisions about policy and the allocation of resources. They reorder priorities. They help foster democracy. Steven Fish, a political scientist at the University of California, Berkeley, argues that democracy is exceedingly rare in countries where there is a large gap in the literacy rates of men and women—and where there are substantially more men than women. The latter suggests that boys are more valued, and receive better nutrition and health care. And as Amartya Sen says, the more men, the more crime and the more violence.

Lessons from the Jungle

Following the Rwandan genocide, a Ugandan farmer was hired to haul bodies out of a lake. Over time, he retrieved so many corpses, wrapping them in plastic sheets and stacking them in piles, that he became numb to the horror. Only one memory really haunted him. "One time I found a woman," he said. "She had five children tied to her. One on each arm. One on each leg. And one on her back. She had no wounds." Who could even dream up such a breathtakingly cruel act?

It was probably a man. But women, too, are capable of some horrific acts. Who can forget Andrea Yates, the psychotic Houston woman who drowned her five young children in the bathtub?

Or Susan Smith, the South Carolina divorcée who coldly strapped her two little boys into their car seats and then rolled the car into a lake so they wouldn't come between her and a would-be boyfriend? Both stories were widely publicized, in no small measure because they were so shocking, because the betrayal cut so deep. Women—mothers—don't do that. Except, of course, when they do.

Still, men are the primary perpetrators of murder, mayhem, and other assorted destructive acts. The patterns are remarkably consistent across time and culture. According to FBI statistics, an American man "is about nine times as likely as a woman to commit murder, seventy-eight times as likely to commit forcible rape, ten times as likely to commit armed robbery and almost six and a half times as likely to commit aggravated assault." Men are also vastly more likely to commit fraud, steal a car, vandalize, commit arson, or get busted for drugs. In fact, American females lead males in only two crime categories: Adolescent girls are a little more likely to get arrested for running away from home, and women are twice as likely to get arrested for prostitution.

Not only are crimes, especially violent crimes, more likely to be committed by males, but the overwhelming majority are committed by young males, those between the ages of fifteen and thirty. To the mother of a boy heading eventually toward that age range, this is not good news.

But do men commit more crimes because they're stronger? Is violence a vestige of physique, not temperament? To answer that question, researchers looked at instances of same-sex murders, where either women killed women or men killed men—theoretically limiting instances where the perp was substantially stronger than the victim. After gathering data from dozens of countries and regions across multiple time periods, they concluded that the "probability that a same-sex murder has been committed by a man, not a woman, ranges from 92 to 100 percent."

Of course, the debate still rages: Are these differences innate? Are males "naturally" more violent than females? Or is the behavior learned—created, passed on, and reinforced by culture? While nurture is surely a factor, nature seems to play an outsize role. Again, the patterns are remarkably consistent. And animal studies—particularly of our closest cousins, chimpanzees—find some interesting similarities between their behavior and ours.

In 1960, Jane Goodall, the renowned primatologist and environmental activist, went into the Gombe National Park in Tanzania to observe chimpanzees, up close and in the wild—something that had never been done. What she saw there rocked the world: She discovered that chimpanzees engage in certain behaviors previously thought to be exclusively human. One chimp repeatedly scooped termites from a mound using a stick—a "tool," she soon realized. And she found that the animals were capable of "reasoned thought, abstraction, generalization, symbolic representation, and concept of self."

But some of her findings were less romantic.

When Goodall first arrived, the thirty or so chimps she was studying lived together in a single group. But over a period of several years, the group split into two separate factions, one occupying the northern part of the range, the other, the southern part. Bit by bit, relations between them went from friendly, to hostile, to murderous. And the violence wasn't just about protecting turf or clan or resources. Groups of four or five or six chimps from one gang would sneak into their rivals' territory looking for a fight. And if they found an individual who was alone or otherwise unable to defend himself (or occasionally herself), they would attack—biting, hitting, kicking, scratching, and dragging their rival to death. "It was only one female who really participated," Goodall told me. "And she was the one who never had babies." The raids continued until the northern gang had wiped out the southern gang, killing all the males, many of the females, and several of the infants. The females who were left were then forced to join their erstwhile foes.

"It was a very primitive war," Goodall recalled. "It's got some of the precursors of human war. It shows you the kind of tendencies we have brought with us throughout evolution that led to modern war."

Chimpanzees and humans are intensely social creatures. Both live in hierarchies where status is important. Both compete to better their position in the pecking order. And both build coalitions to get what they want. But predictably, males and females approach these relationships differently. According to Richard Wrangham and Dale Peterson, authors of *Demonic Males: Apes and the Origins of Human Violence*, a male chimp in his prime virtually "organizes his whole life around issues of rank. His attempts to achieve and then maintain alpha status are cunning, persistent, energetic, and time consuming. They affect whom he travels with, whom he grooms, where he glances, how often he scratches." Female chimps build alliances, too. But they work with other females to whom they feel some emotional attachment. In other words, males get together as a means to an end; females make friends. Sound familiar?

Humans and chimps have something else in common: Of the more than 4,000 mammals and 10 million other species, they are the only two who live in "male-bonded, patrilineal communities" in which groups of (mostly) males raid their neighbors' territories, looking for vulnerable peers to attack and kill. Most animals don't kill their own kind. And when they do, males usually kill infants sired by other males, in hopes of mating with their mothers. (A charming practice that is often successful.)

"Male chimps are very aggressive. Female chimps are aggressive, too, but they can't show it in the same way," Goodall says. "One, they are not as strong. Two, for most of their lives they are protecting a child. So you can't go swaggering around and waving branches and displaying because it puts your child at risk. It's nonadaptive in an evolutionary sense."

DIFFERENT RULES

As the political economist and philosopher Francis Fukuyama says, there is "something to the contention of many feminists that phenomena like aggression, violence, war, and intense competition for dominance in a status hierarchy is more closely associated with men than women." A world run by women would follow different rules, Fukuyama surmises. "And it is toward that sort of world that all postindustrial societies are moving. As women gain power in these countries, the latter should become less aggressive, adventurous, competitive, and violent."

Hear! Hear! But Fukuyama goes on to note a potential flaw in this thinking: If the male predisposition toward violence, power, and status is indeed rooted in biology, it is harder to change—in individuals and in societies. "What is bred in the bone cannot be altered easily by changes in culture and ideology." True enough. But the idea that certain tendencies are "bred in the bone" works both ways. If men aren't easily made *less* violent, then isn't it also true that women aren't easily made *more* violent? And doesn't the "Margaret Thatcher argument"—that as women gain power they just act like men—start to collapse?

Still, if male aggression (particularly among the fifteen- to thirty-year-old set) isn't going to change—or at least not much—it will have to be controlled. Often, that's been accomplished by directing it "outside the community"; wars abroad have long been used to help foster tranquillity at home. But modern society has created other options, more peaceful options. To begin with, a "web of norms, laws, agreements, contracts, and the like" that are a staple of liberal democracies can help restrain some of that surging testosterone. And the proliferation of hierarchies, as it were, can help channel aggression into productive activity. No longer is proving your bona fides on the battle field the only way to improve your social standing back home; now a guy roiling with competitive juices can start an Internet company, win a tennis match or an election, write a screenplay, or find new ways to distribute life-saving drugs to rural villages in Africa.

Meanwhile, women have to keep pushing their way into the arena, taking on roles such as voter, community activist, legislator, executive, diplomat, prime minister, president. In many countries and in many parts of the world, that's already happening. And women are bringing a different perspective to foreign policy and national security issues. For example, American women are less supportive of U.S. involvement in war, and women in general are less likely than men to see force as a legitimate tool for resolving conflicts. While stopping short of identifying causation, Fukuyama says that "increasing female political participation will probably make the United States and other liberal democracies less inclined to use power around the world as freely as they have in the past."

In developed democracies, women have substantially more power than they do in authoritarian states. Just scanning the

Inter-Parliamentary Union's annual ranking of countries with the most women in their national legislatures provides anecdotal confirmation: After Rwanda, Sweden, Finland, Costa Rica, and Norway lead the way. Clustered at the bottom? Those bastions of freedom and enlightenment: Saudi Arabia, Qatar, Kyrgyzstan, Yemen, and Egypt.

Academics continue to argue about cause and effect: Do democracies create more opportunities for women—or does empowering women strengthen democracies? Regardless, democracies tend not to go to war against each other, which is why supporting governments that are democratic (or aspire to be) is a long-standing tenet of American foreign policy. But what about the parts of the world that are less developed, less democratic? Where the Saddam Husseins and the Mahmoud Ahmadinejads still reign, seeking to subjugate their people and dominate their neighbors, often by acquiring weapons of mass destruction? Can women be trusted to deal with these despots? Here, Fukuyama argues, "masculine policies will still be required, though not necessarily masculine leaders."

Perhaps. It is true that the world's most powerful women have often pursued "masculine" policies. Margaret Thatcher went to war in the Falklands. Indira Gandhi oversaw the development of nuclear weapons. And Queen Elizabeth I had her rival, Mary Queen of Scots, executed.

What about Hillary Clinton, the most successful female presidential candidate in American history? It seems that not just her policies but often her approach also tilt masculine. And that may help explain why voters have yet to fall in love with her: In a world where people have different expectations for women, she pays a price for showing us her steel spine more often than her soft heart. But no one questions her toughness. And in this dangerous and uncertain world, toughness is the indispensable quality for any would-be president.

That said, different kinds of women leaders have succeeded in less militarized, if no less masculine, cultures. Take Ireland. First Mary Robinson, and then Mary McAleese, was elected president. A Roman Catholic born in Northern Ireland, McAleese was at first considered divisive; an opposition journalist once referred to her as a "tribal ticking time bomb." But she has reached out to both sides in the long-standing and bitter clash between Catholics and Protestants—and both sides have embraced her. McAleese says "building bridges" is the theme of her presidency, and when she ran for a second seven-year term in 2004, she was unopposed.

THE PEACEFUL APE

If our family tree suggests that violence in humans is innate, it may also suggest a strategy for controlling that violence. Bonobos, the "peaceful apes," look very much like chimpanzees. While they're a little bit smaller, they share so many physical characteristics with chimps that scientists didn't recognize them as separate species until the 1980s. But when it comes to their social lives, bonobos are a world apart. As Wrangham and Peterson explain, they have forged a "threefold path to peace. They have reduced the level of violence in relations between the sexes, in relations among males, and in relations between communities." How?

First, the sexes are "codominant." In chimpanzee society, every male is dominant to every female. And he lords it over her: If a female chimp fails to offer appropriately submissive gestures to a higher-ranking male—like crouching down and panting in his presence—she gets a brutal beating. Bonobos also establish hierarchies. But the alpha female is equal to the alpha male, just as the lowest-ranking female is equal to the lowest-ranking male. "In between," Wrangham and Peterson say, "your rank depends on who you are, not what sex you are."

In addition, female bonobos cooperate with each other—they build relationships—in ways that males don't. And they use their collective power to protect themselves and their offspring, particularly their sons. (Female bonobos, like female chimps, leave their troops at adolescence to guard against inbreeding, but the males stay with their mothers.) If another male threatens a female or her son, mama calls for help, and her posse of girlfriends rallies to her and her son's defense. When these all-girl gangs get going, they can defeat even the highest-ranking males. Female bonobos will also use their collective power to get their fair share of certain resources, like favorite fruits or meat. Male chimps always get first dibs on the good stuff. But once again, female bonobos work together to keep male aggression in check—even though male bonobos, like male chimps, are roughly half again as large as females.

Finally, bonobos don't stage murderous raids on other groups of bonobos, as the chimps in the Gombe did. While smaller groups often avoid larger ones and fights do break out, meetings between neighboring bonobo communities can be peaceful, even friendly. And the friendliness is *always* initiated by the females.

In the end, all of these behavioral differences add up to a single, inescapable conclusion: Male bonobos are less violent than male chimpanzees. And the reason? Girl Power.

Perhaps it's wishful thinking, but maybe there are some hopeful notes here for humans. While male violence—and the dominance of men over women—has long been with us, perhaps bonobos teach us that neither is inevitable. And that female power isn't just a mirror image of male power. Women have certainly been known to abuse power: As prime minister of India, Indira Gandhi used emergency provisions to grant herself extraordinary powers and quash dissent. But nature and history suggest that human females—like bonobos—also use their power to protect themselves and their families; to resolve conflicts; and to build lasting relationships. All of which can help build and sustain peace. In Rwanda and in Bosnia. In Northern Ireland and in South Africa. In established democracies and countries trying to create a better future.

That's not to say women should replace men altogether. Plenty of men have long been both committed to peace and heroic in defense of others, whether on the battlefield or in the stairwells of the World Trade Center. Still, it seems clear that increasing the numbers and influence of women, allowing them not just to speak but to be heard, would make the world more peaceful.

"Clearly, I do believe we have inherited aggressive tendencies—you can't look around the world and say that we don't have some kind of innate response to certain actions which is an aggressive response," says Jane Goodall. "But at the same time, chimpanzees show love, compassion, and altruism; we have inherited those things, too. And with our huge brains, we have far more ability to control our genetic inheritance than any other creature. War is not inevitable. Human aggression can be controlled."

And maybe, just maybe, it's women—like our bonobo sisters—who hold the key.

Getting to Win-Win

"It may be the cock that crows, but it's the hen that lays the egg."

—Margaret Thatcher

Alexis Herman, the U.S. secretary of labor, had an ominous feeling. Contract negotiations between United Parcel Service, the nation's largest package delivery company, and the International Brotherhood of Teamsters, the union representing tens of thousands of UPS workers, were bogged down. If the two sides couldn't work things out before the deadline—midnight on August 4, 1997—the workers would strike.

Herman had been sworn in as secretary less than three months earlier, after a bruising confirmation process. Questions about her qualifications—as well as her ties to political donors—had left Congress uneasy and the labor movement unenthusiastic. She was still learning her away around many of the department's complex issues, including the status of the Teamsters-UPS talks, but her staff kept assuring her the deal would get done. "They said, 'Oh, don't worry, Madam Secretary. Don't worry. We're going to get down to midnight and they're not going to walk,'" she recalled. But Herman was skeptical. "I had this gut, okay. Call it women's intuition, call it what you want. And I kept thinking, 'Lord, don't let me start my tenure off this way.'"

When the midnight deadline passed, the phone rang in Herman's Chicago hotel room: There was no deal. One hundred eighty-three thousand workers walked off the job, kicking off the biggest strike in the United States in more than a decade. During a week-long cooling off period, federal mediators tried—unsuccessfully—to restart the talks. By Saturday, August 9, both sides had dug in for a protracted fight.

Meanwhile, the economic impact of the strike was growing. UPS handled more than 80 percent of the nation's ground shipments, delivering 12 million packages a day and accounting for roughly 5 percent of the nation's GDP. Already, three governors and thirty-seven CEOs—including those from Sears, JCPenney, Toys "R" Us, and the U.S. Chamber of Commerce—had called on President Clinton to intervene and order the strikers back to work. He refused. But he asked Herman to take the lead in trying to jumpstart negotiations.

Her staff advised her to keep her distance, to give the situation more time to play out before getting directly involved. But Herman knew instinctively she didn't have that luxury. "Well, I kept saying, 'I'm damned if I do and damned if I don't,'" she told me. "I'm going to get the blame if this thing goes down. And if it gets settled a month from now, nobody's going to give me credit for that after the fact. I said, 'I guess somebody might be able to come in and make nice and be celebratory, but I don't think I have the same degrees of freedom as the average, previous, *male* Labor secretary.' We [women] are not allowed to fail."

So Herman decided to forge her own path forward. "I said, 'It may be a risk, but it's a risk I've got to take. It won't be the first time, and if I go down, I go down trying.'"

Over the next several days, she engaged in what she described as "shuttle diplomacy," staying in constant contact with both sides, including a series of phone calls from Air Force One during a trip to St. Louis with the president. Within days, she had convinced UPS chairman Jim Kelly and Teamsters' boss Ron Carey to come back to the table. "I didn't play games. I felt it was important to create a feeling of honesty in the process. And I worked hard at building that trust." Among other things, Herman promised that when the sides met, she would stay with them until the deal was done. "And again, that was a huge risk because everyone was saying, 'How do you know you can get a deal done?' I said, 'I *don't*.'"

Rather than holding talks at the Department of Labor, Herman invited the teams to meet at the Hyatt Regency Hotel on Capitol Hill in Washington, DC, where they could eat, shower, and hopefully sleep for a few hours each night. With the sides far apart, talks began inauspiciously in the Grand Canyon Room. Over the next four days, working nearly around the clock, negotiators made steady progress. Then the talks bogged down again. Even though both sides had agreed not to talk to the media during the process, Herman got word that the Teamsters were planning to leave the hotel to attend a press conference and rally nearby. "It was kind of like an in-your-face move, and it really put my credibility on the line," she said.

"Then someone came and told me the door they were going to sneak out of. I think it might have been one of the waiters," she recalls with a laugh. So Herman—elegant, polite, impeccably dressed and all of five feet three inches tall—planted herself in front of the exit and waited. Soon enough, a group of Teamsters came down the hallway, led by one of the negotiators. "And I said, 'You're not doing this.' And he said something; he might have called me 'little lady,' I can't remember what," she says, her voice growing soft. "I got up and I grabbed him by the lapels . . . and I just looked at him and I said," she is almost whispering now, "'Don't fuck with me.'

"And then I just got up on all my heels and I said, 'You want a press conference, you've got a press conference. If you want these negotiations to break down, you will have to break them down because I won't.' And he looked at me and I looked at him, and we just sort of did that for a while. Then he turned around and went back. And then, over the next twenty-four hours, we got it done."

It had been a baptism by fire. But even before UPS's trademark brown trucks were rolling again, Herman was being heaped with high praise from all sides—even from some of those who had questioned her qualifications at the outset.

John Calhoun Wells, the nation's chief labor mediator who had not met Herman before the strike, said she was "absolutely crucial" to the settlement. "She's not flamboyant," he said. "She never tried to pressure people with any sort of arm-twisting. Hers was a more subtle style of persuasion—never asking people to do what they perceived to be against their best interest, but suggesting that maybe it was in their interest to give a little to get a little back in return."

Union officials also had strong words of praise. "She was not our first choice," acknowledged Gary Shea of the AFL-CIO. But he rated her performance "at least a 9, if not a 9.5." People who have seen her up close, he added, "both publicly and privately, as we have, are extremely impressed by her presence. There's a grace she has. It's the key to why she is so good at the interpersonal stuff."

I admit, I love this story. The image of Alexis getting "up on all [her] heels," as she tells it (which would have made her, oh, about five-six) to have it out with some (one imagines) burly labor guy, is just so satisfying. Little lady *that*, mister! But in truth, it's not just that she surprises some people by being tougher than they expected. It's that she sort of sneaks up on them by being *better* than they expected. She got the job done not by threatening or posturing or double-dealing, but by helping both sides find win-win solutions, by leading them to common ground. It's a style that is by no means unique to women. But it is one for which a lot of women—like Alexis Herman—are well suited.

A Matter of Style

All of which begs the question: Do men and women lead differently? Is there such a thing as a "female style"? A recent analysis of forty-five separate studies addressing the question found that the answer was "yes." Women are slightly more likely to be "transformational" leaders, collectively setting goals and empowering their teams to achieve them. And men are more likely to be "transactional" leaders, letting subordinates know what is expected, rewarding them for their successes and holding them accountable for their failures. Not surprisingly, most leaders did not fit neatly into one or the other of these categories, but there was, nonetheless, a measurable difference based on gender.

Now comes the kicker: Research also shows that transformational leaders—especially those who also reward good performance, a positive aspect of transactional leadership common among women—tend to be more effective, particularly in the less-hierarchical, fast-paced, and innovation-driven contemporary world. So not only do women have a somewhat different style; it's more likely to be successful.

To me, what's most important about that finding is not that women rule (though I obviously have a soft spot for studies and statistics that put us girls in the most flattering light). Rather, it's further evidence that there is more than one way to bring home the bacon and fry it up in a pan, that different leadership styles—regardless of their gender bent—can get the job done. And that gives everyone more options; it creates a more flexible, more adaptive and ultimately more productive workplace.

"By valuing a diversity of leadership styles, organizations will find the strength and flexibility to survive in a highly competitive, increasingly diverse economic environment," says Dr. Judith Rosener of the University of California, Irvine.

Sally Helgesen, a leadership development expert, believes that because women have rarely fitted easily into corporate molds not designed for them, they have been "forced to pioneer policies and strategies that are turning out to be exactly suited to the conditions of the new knowledge-based economy. In the end, women's greatest contribution to our changing world may be their insistence upon breaking the mold rather than just fitting in."

Among other things, the line between work and home is fading, and people—especially women—are learning to invent their own positions. I more or less invented my current "job," which I sometimes describe as "stay-at-home pundit." It's an interesting and flexible mix that has included contributing to

Vanity Fair, giving speeches, yakking about politics on television, consulting on politically-themed movies and television shows, and writing about stuff that interests me. I work out of an office in my house, which saves me time spent commuting (and I confess, on some days, showering). My children have (mostly) learned to respect my closed door, and when they don't, I escape to the local public library, conveniently equipped with free wireless Internet. The technological innovations and cultural transformations that allow me to do what I do came together just in time for me. While I realize that it can't work for everyone, there's no question that opportunities to define a career path will continue to increase—a trend that I believe will be led by women.

The biggest downside to my current arrangement is the anxiety I feel when I face the "occupation" line on a school form or loan application. I usually write "consultant"—and then hope I don't get busted for I'm-not-sure-what. There's also a certain guilt that comes from not having to leap out of bed before dawn to unload the dishwasher, fold the laundry, shower and blow-dry and apply makeup, get the kids ready for school, and burn rubber backing out of the driveway at 7:45 a.m. I recently saw a cartoon that summed up my life. A couple is sitting at the kitchen table in their bathrobes, drinking coffee. As the man taps away on his laptop, his wife says: "You've blurred the boundary between working from home and being unemployed."

This increasingly less structured, more flexible workplace suits women's lives—and their skills. "When you put together a thirty-person project team [in the past], it was all people from Raytheon," explains Tom Peters, the management consultant. "Now, the thirty-person project team involves people from eleven companies, seven countries, and three continents. There's no formal power or hierarchy. So we need a different set of relation-driven skills."

"This is why you want to hire women," says Pat Mitchell, a pioneer in broadcast news and the current president of the Museum of Radio and Television. "They are consensus builders. They really do look for different ways to resolve things. They are innovative and creative thinkers. And they do act on instinct and intuition."

TRUSTING YOUR GUT

When Kathleen Sebelius was insurance commissioner of Kansas, Anthem, a massive health insurance company, proposed buying Blue Cross Blue Shield, something it had done in seven other states. Anthem had already received approval from Blue Cross Blue Shield's Kansas shareholders, after they'd been assured the deal would be profitable. But Sebelius, who had final say, was skeptical. Blue Cross Blue Shield wrote policies for nearly two out of three covered Kansans, making it by far the state's largest health insurance provider. "My gut just told me this was a disaster waiting to happen, that we did not want to be acquired by an out-of-state, for-profit company, that we did not want to lose that level of control."

Doctors and hospitals were vocally opposed, but Sebelius wanted to hear from a broader range of people affected by the takeover. So she set up a series of meetings across the state. By the time she got to the first one, it was clear that the issue had "touched a nerve," she told me. "It was a freezing cold day early in the morning, and there were 350 people in this room and more lining up outside. I mean, there was no organized anything, except we put out the word and told the medical society that I'd be coming and that they should tell other people. I mean, the room was totally packed, and people were up in arms."

After a series of similar meetings around the state, Sebelius held an administrative hearing—a trial, as it were. Over the course of five days, both sides argued their case, with the insurance commissioner as judge. "At one point, the heads of the Anthem Company were all seated in these folding chairs in this ballroom. And these guys are all making twenty-one million bucks, or thirty-four million bucks. And I said to somebody when we took a break, 'I think they think they died and went to hell.' They never in their wildest dreams expected to be in Topeka, Kansas with this girl up there . . . They were so sure it was going to be successful that they'd already printed stationery."

After refusing to let her political consultants take a poll, Sebelius weighed the evidence—and blocked the takeover. "First time in the country that had happened," she said. "I was sued by the company, but the Supreme Court upheld my authority to do that. And it was a hugely popular decision around Kansas, kind of the classic standing up for the people against the big guys." It sprung from a gut feeling—and it paved the way for her election as governor the following year.

As more women assume leadership positions, they're increasingly talking about the role of intuition in their decision making. "All the women leaders I've met led with a greater sense of intuition than men," Oprah Winfrey once said. "I am almost completely intuitive. The only times I've made a bad business decision is when I didn't follow my instinct. My favorite phrase is, 'Let me pray on it.' Sometimes, I literally do pray, but sometimes, I just wait to see if I wake up and feel the same way in the morning. For me, doubt normally means don't. Doubt means do nothing until you know what to do. And I'm really, really attuned to that."

Dianne Feinstein agrees. "I do think there's such a thing as intuition. And I do believe that women bring a certain intuition about the human dilemma, the human problem, the sense of priorities. All of that."

That intuition, according to some, makes women better decision makers. "I think we make them faster," says Jane Friedman, CEO of HarperCollins, the publisher of this book. "I don't think men trust their instincts. We are instinctual. That also comes from being child-bearing. We have the gut, the nose, whatever you want to call it."

As history would have it, men are logical, women intuitive. And guess which one is considered superior? Ding, ding, ding, you win the toaster! Logic trumps that touchy-feely intuition every time. Girls are taught from the earliest age to doubt that feeling in their guts. I can't remember a particular example of a teacher or parent or coach pooh-poohing intuition per se, but the very strong message was: Intuition is just another form of—dare we say it—emotion! And emotion must be stamped out, replaced by orderly reason.

That's not to say that intuition is always right; it's not. I was convinced my first child was a boy until the doctor put her in my arms. At the same time, I've never actually seen proof that men are more logical. But the anecdotal evidence suggesting that people believe it is as thick as the barnacles on the bottom of an old boat. President Richard M. Nixon—he that bastion of equilibrium—considered appointing a woman to the Supreme Court in 1971 because he thought it would help him politically. But that didn't mean he thought it was a good idea. "I'm not for a woman in any government job, whatsoever, mainly because they are erratic and emotional," he told attorney general John N. Mitchell, in a conversation immortalized by the president's secret tape machine. "Men are erratic and emotional, too, but the point is a woman is more likely to be." Those crazy broads. Next thing you know, they'll be talking to portraits.

Because men don't necessarily experience intuition the way women do, they seem to believe it springs from that same mysterious corner of the female imagination as PMS and weird cravings during pregnancy. But new studies are finding intuition actually has physiological roots. "Gut feelings are not just free-floating emotional states but actual physical sensations that convey meaning to certain areas in the brain," explains Dr. Louann Brizendine, author of *The Female Brain*. "The areas of the brain that track gut feelings are larger and more sensitive in the female brain, according to brain scan studies. Therefore, the relationship between a woman's gut feelings and her intuitive hunches is grounded in biology."

Among other things, women are particularly sensitive to what other people may be feeling. An analysis of 125 separate studies on the topic by Judith Hall, a professor of psychology at Northwestern University, found that "women are better than men at decoding emotional messages—and better at spotting lies."

Men, on the other hand, are much more likely to miss the signals—especially signs of despair or distress. "It's only when men actually see tears that they realize, viscerally, that something's wrong," Brizendine explains. "Perhaps that's why women evolved to cry four times more easily than men—displaying an unmistakable sign of sadness and suffering that men can't ignore."

I versus We

After I left the White House, I joined CNBC as cohost of the left-right political talk show *Equal Time*. My partner, the inimitable Mary Matalin, and I disagreed on most things ideological; still, we had a lot in common. We were both raised in Catholic families without any political ties, and neither of us much imagined we'd end up in Washington. We were both crazy for her husband, James Carville, my colleague from the Planet Clinton. And we both wanted our show to be more of a dinner party than a food fight. And so it was. We had an unspoken agreement that we would disagree respectfully—and, whenever possible, with humor. It didn't hurt that I thought everything Mary said was funny; when she would cleverly lampoon me or one of my Democratic fellow travelers, I was often laughing too hard to respond effectively. But we also listened to each other, and—*quelle horreur!*—we sometimes agreed. It was a good time, for us and for our guests, and Mary and I became great friends.

It didn't start out that way. As Mary liked to tell it, the first time she ever heard of me was when she saw me on CNN, calling for her to be fired. It was the early stages of the 1992 presidential campaign, and my boss, the upstart governor of Arkansas, was taking on her boss, who was then living in the White House. At some point, Mary called Governor Clinton a "draft-dodging, pot-smoking, womanizer"—a real gift for understatement, that Mary. So I told reporters that President Bush should fire her. Carville told me that was a dumb move, that it only dragged the story out and gave the media a chance to repeat Mary's irresistible (and damning) line. Of course, he was right. The press played the story all day, and the president didn't fire anyone.

After the campaign, Mary famously married James, and she and I occasionally saw each other socially. Our inauspicious beginning was never an issue. In fact, Mary was extremely helpful to me when I was trying to figure out exactly how and when to leave the White House, something for which I will always be grateful. And it was she, along with then–CNBC chief Roger Ailes, who convinced me to come to *Equal Time*. (Ailes is now the president of the Fox News Channel. And despite our political differences, he was one of the best bosses I've ever had.)

It wasn't that Mary and I weren't competitive; we both wanted to make our points, to win the argument, as it were. But we didn't want to do it by crushing each other. We believed we had more to gain from a partnership than from a duel to the death; we always wanted the terms to be win-win. Was it because we were women? It seems that's at least part of the answer.

First, women are all about protecting "the relationship." According to Brizendine, girls' (and women's) brains are wired to get what they want. And what do they want? "To forge connection, to create community, and to organize and orchestrate a girl's world so that she's at the center of it. This is where the female brain's aggression plays out—it protects what's important to it, which is always, inevitably, relationship."

Or as Tom Peters recounted, a woman he knew was about to start writing a newspaper column. "The advice from the men was: Never sacrifice a good column for a friendship," he told me over breakfast one morning. "The advice from the women was: Never sacrifice a friendship for a good column."

But in addition to protecting relationships, the win-win mentality is also about results. "I do think women approach things—at least some women—differently," says Governor Sebelius. "You've got to get everybody at the table. You've got to pay attention to what they're saying and not just pound them into the sand if they're not doing things that you want them to do."

You've also got to give a little to get a little. "Always leave a little something on the table," says Susan Lyne, president and CEO of Martha Stewart Living Omnimedia. "It's important advice in any business. A total win for one side in any negotiation is just wrong because it's almost always a Pyrrhic victory. You end up with bad partnerships."

If I had a dollar for every meeting I've sat through, listening to some man take credit for my work, I wouldn't need to buy lottery tickets. It's not that men don't often deserve some of the kudos; of course they do. It's just that they tend to take more than their share. They always seem to go right for the biggest brownie on the plate. What's more, they expect the other guys (and sometimes gals) to do the same.

Governor Sebelius said that her willingness to work collaboratively with (mostly male) legislators on important bills, including health-care reform, was key to getting them passed. "We didn't say, 'No, it's got to be my baby.'" Rather, she said, the focus was on the results. "How do we get this passed? How do we keep people at the table? How do we get our ideas—if they're not there yet—into the process? And it's probably letting them—the boys—have their name on the bill and pat them and stroke them and tell them, 'Oh, you're so brilliant. Oh my god, you're so brilliant. Why didn't I think of that?'"

Women don't just share the things that make them look good; they're also more likely to tell you the stuff that doesn't make them look good. "Women are more honest about their shortcomings," says Debra Lee of BET. "I think men know it and hide it. I hope they know it. But women will be honest. And that's why I think they're good managers, because they hire people that have skills that they don't have. Men are just like, 'I can do it all.' I think women really reach out.

"But I see in some women that creates a hesitancy in maybe going for the next position, or promoting themselves, or speaking out. And men just talk. They just like to hear themselves talk whether they really have something to say or not. And I see with women, a lot, that they just wait until they really have something to say, and really have something to add. They're not going to sit there and just talk to impress everyone. It's a rare woman that you see do that; I look around meetings all the time and notice that."

Communication, Collaboration, Consensus

When you don't care who gets credit, it's easier to work together. Just ask San Francisco's top public safety officials. They call themselves "the Sirens," but the city's residents call them in an emergency. And the fire chief, the police chief, the district attorney, the director of emergency services and homeland security, the coroner, and the medical examiner are *all* women. Except for the district attorney, who was elected, they were all appointed by mayor Gavin Newsom. "I wasn't looking for the 'woman' candidate; I was looking for a competent team," he said. He got it. And in the post–Hurricane Katrina world, it's more important than ever. "America loves the macho guy with the cigar and crew cut. But America also likes results. I've often sat in envy of the ability of women to multitask, put ego aside, not complain and solve the problem."

One of the hallmarks of this all-girl group is collaboration. District attorney Kamala Harris and police chief Heather Fong inherited departments that were at each other's throats. (In fact, the previous DA actually indicted the police chief and some of his top commanders following an altercation that involved a bag of leftover fajitas. Really.) Unlike their male predecessors who spent years fighting—but not speaking—the two women have managed to keep the lines of communications open, even when they disagree. And they've looked for new ways to solve existing problems. "We have to dispense with the old conversation about being 'soft' on crime or being 'hard' on crime; we have to talk about being 'smart' on crime," Harris says.

"There are many ways to mediate and defuse situations," explains Fong. Among other things, she encourages officers to get out of their patrol cars and onto the street, where they can meet the neighbors. And she has teamed up with fire chief Joanne Hayes-White to conduct joint training for police officers and firefighters, who often respond to the same emergencies.

Tactics that aren't "badge heavy" are common in other large cities around the country, such as Boston, Milwaukee, and Detroit, where women are in charge, says Maggie Moore, director of the National Center for Women and Policing in Arlington, Virginia. "Eighty percent of modern policing is about communication, prevention, and management," she says.

And while women don't have the market cornered on these qualities, they do tend to bring a certain perspective to their roles as leaders. "I do think mostly women are more collaborative," says Senator Kay Bailey Hutchison. "I do see that. And I think that women being in leading positions in business and in the legislative arena have influenced the whole system that we're in. I think corporate governance is much, much influenced by women being in high echelons now. There is a more collaborative business model. Same with government, with the legislative arena. Women have gone to the top fast."

Susan Hockfield, the first female president of the Massachusetts Institute of Technology, is a case in point: She says she would like the university she leads to work together in new ways.

"I've been excited by an interdisciplinary approach since I was a student because it brings together knowledge that no one comes to on his or her own," she said.

In discussing how she would approach being president, Senator Hillary Clinton said it was very important for a president to gather information from a wide variety of sources. "I seek out people who are not only able to come with some expertise or relevant experience, but are willing to debate and discuss differences of opinion," she told *Newsweek*. "Sometimes it surprises people to see how seriously I seek out that kind of debate. Obviously, I can't know every nook and cranny of what a decision might mean. I want people to try to reach a consensus, but if a consensus is not easily available, I want to know all sides of an issue."

Defining Power

With leadership comes power. And for generations, people have asked: Do women lack power because they don't want it enough? I confess, I don't know the answer. For starters, I've found that a lot of women are uncomfortable with power, at least as traditionally defined. Most of the women I interviewed for this book shrank from the notion that they were "powerful." Influential? Perhaps. Powerful? Don't go there.

Contrast that to a meeting I once had with Michael Ovitz, who was then head of the Creative Artists Agency and widely considered the most powerful agent in Hollywood. "Power is like a club," he said, gesturing toward an imaginary weapon lying on the floor beside his chair. "As soon as you reach for it, you lose it." The strong implication was: Michael Ovitz's club never leaves the rug!

Most women don't think like that. "They don't like power," the late Anita Roddick, environmental activist and founder of The Body Shop, told me. "They see what it's done to men and they want no bloody part of it. They like moral influence. They like the ability to change things. But in terms of how we define power now? I don't think I see it, you know? Women are moving away from the big institutions in numbers, and they are joining smaller groups. And so I think until women want to reclaim the notion of power, I think it's going to be quite hard."

True, more and more women are forgoing those "big institutions" and looking for professional opportunities that accommodate their values and the demands on their time. Surely that is one of the reasons that women-owned small business continues to be the fastest growing segment of the economy. But that doesn't mean women don't want to play a role on the bigger stage as well.

"This is what women know: They know any of the great social justice movements emerged and grew up out of a sense of shared community," said Roddick. "The women's movement, the gay rights movement, the ecology movement. They have become mass movements, and women are responsible for making them flourish. And I think it's women's sense of consciousness raising that has got this business off the ground. And I don't think we should ever, ever lose that sense of shared, collective action. This is our heritage. This is the story that we've written. Rather than this notion of power, this is it."

As more and more women find themselves in positions of power, they're finding new metaphors to describe their world. For men, it's always been sports and war. For women, increasingly it's motherhood. Again, not all men like sports. (My husband, for one, is infinitely more interested in cooks than quarterbacks.)

Moreover, not all women are mothers, nor do all of those who are believe that raising kids alone qualifies them to rule the world. That said, motherhood does build a broad range of valuable skills. Moms learn to anticipate needs. (How long can we go before someone has to "go"?) They learn tough decision making. (When are the kids ready to ride their bikes to the park by themselves?) They learn diplomacy. (How do you get three kids to agree on one TV show?) And they learn discretion. (What happens in the minivan stays in the minivan.) They also learn sacrifice, time management, multitasking, hard work, long hours, flexibility, and team building. Who wouldn't want leaders who had honed these skills in the tough crucible of real life?

A 2001 study by the Center for Research on Women at Wellesley College found that highly successful women leaders are increasingly speaking of mothering as both a training ground for leadership and a metaphor for leadership behavior. "It's a sign of their comfort with motherhood," said Sumru Erkut, the study's author. "In the past, women checked their womanhood at the door." But now, rather than treating the time spent managing children as a liability, more and more women are claiming it as an asset. They're retooling the terms of the discussion.

"I think I'm a better leader in my job because I'm a parent," says Eileen Collins, a NASA astronaut and the first woman to command the space shuttle Discovery. "I am a lower-stress person, more organized, and have learned how to set priorities because of my children."

Likewise, Joanne Hayes-White, chief of the San Francisco Fire Department and the mother of three young sons, says being a parent has been the best preparation for leading the city's 1,700 firefighters. "It's about consistent discipline, setting clear boundaries, rules, and expectations." Being a mother has also honed her more nurturing skills, which are also vital to her role as fire chief. When one of her firefighters was injured on the job a few years ago, Hayes-White visited her in the burn unit almost every day until she was released.

Despite the evidence—anecdotal and empirical—not everyone is certain there is a female brand of leadership. "I always find it very difficult to say because women have never been given an opportunity to lead so that we can truly demonstrate that we can provide a different kind of leadership," says Wangari Maathai, the Kenyan Nobel Peace Prize laureate. "And at this time, because we find ourselves in the leadership positions in the structures, in institutions, in a way of thinking already dominated by the way men think, you almost have to think like them in order to enter their little house," she says, laughing her infectious, subversive laugh.

Judith McHale also expressed doubts. "I think that pigeonholes men and women. I think your leadership style changes and evolves with your responsibility and changing conditions. So I wouldn't want to say one size fits all. I think that's too extreme."

Instead, McHale believes that leadership styles may be influenced more by generation than by gender. "I think older male executives are very much command-and-control," she explains. "When I say older, I mean guys who are in their sixties and seventies. Guys in their forties and fifties are probably more collaborative. So I don't necessarily think it's gender. I think it's generational. I think that any organization that you went into now, the command-and-control structure isn't going to work. It's just too difficult to do it in complex organizations."

Labels aside, women bring a multitude of strengths to their roles as leaders. And because most "combine a strong focus on results with equal attention to the growth and development of the people surrounding them," their success creates more options for *all* leaders, regardless of gender. But that doesn't mean women don't face some tests that men don't. Certain tools are off the table for them; certain behaviors simply aren't allowed.

Study after study has shown that people associate certain traits with men, others with women. Words that define the stereotypical male—aggressive, ambitious, assertive, forceful, self-confident—shout *"Leader!"* Words like helpful, kind, friendly, sympathetic, and affectionate may make people think of their mothers (one hopes), but not their bosses. Moreover, typical male attributes don't always work for women. When women come off as intimidating, fuggedaboutit! It's a one-way ticket off the fast track. And when they try to discipline subordinates? They're perceived as "less effective and less fair than males."

It's hard for women to find just the right balance, to be true to their own natures (whatever those are) without undermining their effectiveness, to be leaders without being men. Eventually, women will have to redefine leadership, redefine power, and then claim their place. And in many ways, that's already happening.

"One of the things I think, you know, when it comes to do the obituary on me, would be I tried to change the language of business," said Anita Roddick, who built The Body Shop into a billion-dollar business without sacrificing her principles, before dying suddenly last year at the age of sixty-four. "I tried to bring in expressions and words that if you teach it becomes constant. You know, I talked about joy in the workplace. I talked about how you measure and develop the spirit. I talked about love." And it worked. RIP, Anita.

Part III

How Women Can Rule the World

Chapter 7

Plugging the Leaky Pipeline

"I have yet to hear a man ask for advice on how to combine marriage and career."

—Gloria Steinem

When I was younger, I didn't see any particular obstacles in my career path. I moved from campaign to campaign, taking on more responsibility as I gained experience. The biggest barrier for a lot of my peers—male and female—was the perpetual lack of job security. Being a Democrat in the 1980s meant losing a lot of races, and finding work after a string of defeats wasn't always easy. But I was single and childless, and the prospect of being unemployed never cost me a minute of sleep. Even at the time, I viewed my relatively high risk tolerance as a significant asset—and my gender as a net neutral.

After a few campaigns, however, I began to notice that there were fewer women standing on every successive step of the career ladder. There had never been many in the inner circle, as far as I could tell. But I expected that would change, as the next generation of women rose through the ranks. Instead, I found that even many of the young women who had started at the bottom with me—organizing volunteers during the Mondale campaign—were gone. Surely, some of them just got tired of electoral politics. The hours are long, the pay is lousy, and you're always looking for your next job. A lot of young men probably left for the same reasons—but there were still enough of them left to fill the top jobs in campaigns and political offices around the country, year in and year out.

And it wasn't just politics. In any number of professions—law, medicine, business—equal or nearly equal numbers of men and women were hired out of college or grad school. But over time, the women seemed to disappear, a trend that has continued over the last couple of decades. Which raises some serious questions: Why are women still so scarce at the highest levels of public life? Even though more and more women are entering the "pipeline"—being hired into entry level positions in a wide variety of fields—why are so few still making it to the top? Why does the pipeline leak?

There's no simple answer. Some of the reasons are obvious: Balancing big jobs with small children is still a challenge. Others are more insidious: Women are locked out of the informal networks where important business gets done. We also have to recognize that men and women often want different things—and that different women want different things, for different reasons, at different times in their lives. And finally, we need to rethink the workplace in response to those differences. Increasingly, employers are realizing that they *can* be more flexible. That they can root out the invisible barriers that poison careers like so much toxic gas. And as employers create new and innovative ways to offer women more choices, and better choices, they're finding a huge reservoir of untapped talent. Plugging the leaky pipeline isn't just the right thing to do, it's the smart thing to do.

Barriers, Seen and Unseen

But first, the obstacles. There's no question that balancing work and family is still particularly hard on women, starting with the lopsided expectations. "Jane Swift had twins while she was governor of Massachusetts; John Engler, who was the governor of Michigan at the same time, had triplets," explains Debbie Walsh of the Center for American Women and Politics at Rutgers. "Nobody ever asked John Engler, 'How are you going to manage being governor while you're raising these little girls?' Yet Jane Swift could not stop getting asked that question."

Regardless of whether they have paying jobs, women still spend significantly more time taking care of the kids and managing the household than the men in their lives. I know I do (although my husband does most of the cooking). While studies show that men have picked up some of the slack in recent years, they're still more likely to be slackers on the home front—even when their wives work. And these additional domestic demands complicate work life even for the most dedicated professionals. While women no longer feel that they have to hide their families, they still fight the realities—and the assumptions—that come with motherhood. One friend recently told me that when one of the guys in her office comes in late, everyone thinks he's been at a business-related meeting, but when she comes in late, those same coworkers assume she's had a child-care crisis. Another friend, a single mom, told me she was in a pickle recently when her company scheduled a huge conference on the same Saturday she was supposed to pick up her daughter from camp. Making matters worse, her bosses asked her to make an important presentation at the conference. When she demurred, a coworker (male, childless) sent around an e-mail to virtually every high-ranking colleague, complaining that one of them would have to fly in early to cover for her during her "mysterious" absence. It wasn't enough that she had to figure out how to balance competing calls on her time; she also had to worry about whether her colleagues would conclude that she wasn't sufficiently dedicated to her job, that her problems were too often their problems.

That women bear a disproportionate share of the responsibility for their families isn't going to make any headlines, even if it is the mother of all obstacles to women's achievement. Surely men can and will continue to do more. But as Susan Estrich once said, "Waiting for the connection between gender and parenting to be broken is waiting for Godot."

Domestic burdens are not the only reason the pipeline leaks. Hidden barriers, particularly at the highest levels, also block women's paths. Studies show that when they quit corporate jobs, women cite lack of opportunity and general dissatisfaction—*not* family responsibility—as the main reasons. Too often, women are not invited to meetings (or alternately interrupted and ignored if they are); left off of distribution lists; excluded from informal networks; or invited to events, like Saturday morning golf, that conflict with other obligations. These "micro-inequities" are like pebbles in the road that, taken together, become boulders.

In 1997, Mike Cook, then president of the accounting and

consulting giant Deloitte & Touche, realized that while the firm was hiring equal numbers of men and women out of college, ten years later most of the women were gone. When he asked his senior managers why that was true, they told him the women had left for family reasons. But Cook questioned that assumption, and asked his team to get more information. So they contacted many of the women who had left in the previous decade and interviewed them. And what did they learn? Some women had, in fact, left for family reasons. But most had more complicated explanations: They left because they felt that as women, their opportunities were too limited. When major accounts came up, women weren't considered for the positions because they required too much travel. Or because they were in male-dominated industries like automotives. Or because the client was an SOB. No one ever asked them about it, the women claimed. The bosses simply made decisions based on what they thought the women wanted—when what they really wanted was more opportunity. It's one thing to work eighty hours a week when you're climbing the proverbial ladder and can see your way all the way to the top rung. It's another thing to bust your hump when you believe you'll be forever stranded somewhere in the middle.

Stereotypes limit opportunity in other ways as well. They lead people to see what they expect to see and to hear what they expect to hear. For hundreds of years, men were considered better musicians than women, and the overwhelmingly male makeup of the world's great orchestras reflected and reinforced that view. Women, it was said, didn't have the strength, stamina, or soul to play difficult pieces written by men, to be played by men. And while there were a few women playing "feminine" instruments like the violin, there were virtually none playing "masculine" instruments like horns. They didn't look right, didn't have the lung capacity, couldn't handle the job, the thinking went. Then a funny thing happened. Musicians began to organize politically, demanding higher salaries, better benefits and protections against being arbitrarily fired. They also fought for—and won—more objective auditions. Soon, musicians began trying out behind screens; the conductors could no longer see whether they were short or tall, cocky or shy, male or female. Over the years, as screens have become common, the number of women in top U.S. orchestras has increased fivefold.

Similarly, a group of psychology professors was asked to evaluate a series of résumés for a tenure track position. Two different résumés were used, one of a "dream" candidate with extraordinary credentials, the other of a successful, but more average candidate. In each case, half the professors received a résumé with a woman's name on it, the other half, a man's. When it came to the extraordinary candidate, there was no bias; everyone wanted to hire the person, regardless of gender. But when it came to the more ordinary candidate, the difference was huge. The professors rated the same teaching experience and research productivity dramatically lower when they thought the candidate was a woman. And when asked whether they would hire the person, 70 percent said yes for the man—but only 45 percent wanted the woman. Interestingly, the gender of the judges didn't affect the findings.

Of course, it's not just the hiring. Once the women are on faculty, the obstacles don't disappear, as Nancy Hopkins, a professor of biology at MIT, famously showed. Although Dr. Hopkins was a star researcher whose lab had produced a number of important breakthroughs in cancer research, her department refused to give her 200 square feet of additional space. After a series of other frustrations, she began comparing notes with other women on the faculty, and through a surreptitious round of research, she found that the women were routinely given lower salaries, less lab and office space, and smaller grants than men with similar rank and experience. Even though more than half of undergraduates—and a big chuck of graduate and doctoral students—were women, all but fifteen of the 209 full professors were men.

When Hopkins presented her findings to Charles M. Vest, the president of MIT, he had a kind of head-slapping, I-could-have-had-a-V8 moment. "I always believed that contemporary gender discrimination within universities was part reality and part perception," he later wrote. "True, but now I understand that reality is by far the greater part of the balance."

Different Incentives, Different Choices

Studies show that women often gravitate toward jobs with fewer and more predictable hours. "If you're in pediatrics, if you're in dermatology, if you're in primary care, and you have a good group practice with you, you can do that," says Bernadine Healy. "If you're a cardiologist—and the patient needs you—you have to go. You get called in the middle of the night. Everybody has to do night duty. You don't get to escape that. And you know there's something about medicine. If you're in one of the tougher specialties, like cardiology, oncology—even if somebody else is taking calls—if one of your patients is in trouble, you don't just say, 'No, I can't talk to you. Talk to my partner who doesn't know you.' So medicine is very demanding that way. But, you know, I have children. And I never failed to take my 3:00 a.m. call."

Women certainly *can* juggle careers, even the most demanding medical specialties, and motherhood. But it's hard. And there's no doubt it discourages some women from even trying. But men are also starting to wonder whether the sacrifices are worth it. "Men are willing to talk about these things in ways that were inconceivable less than ten years ago," says Howard Schultz, the chairman of Starbucks. In a recent *Fortune* magazine survey, 84 percent of men said they'd like more time outside of work, and more than half said they'd be willing to sacrifice income to get it. And so hope springs eternal that someday, work/life balance will be seen as more than just a "women's issue."

Still, lots of my female friends have chosen to downshift or suspend their careers at different points in their lives, for a variety of reasons, in ways that men just don't. I know I did. As my tenure in the White House was winding down and I was thinking about what would come next, I realized that I'd been working eighty hours a week for as long as I could remember. And the first and most important decision I made was that I didn't want to work eighty hours a week anymore. I was thirty-three years old and single, and I wanted to leave time in my life for other things, including, I hoped, a family. I assumed I would keep working—but I wanted fewer hours and more flexibility. My relatively high profile at the White House had left me with lots of options, but before I made any decisions, I wanted to take a little time off. I spent that Christmas with my family, then took a trip to Paris and Morocco with friends.

While I was staying with the American ambassador and his family in Rabat, I got a phone call from Steven Spielberg. He and his partners David Geffen and Jeffrey Katzenberg were starting a new movie studio, he explained; would I come and talk to them about joining their team? So naturally, when I got back from my trip, I flew to Los Angeles for a meeting in Spielberg's office on the Universal Studios lot. After two hours where we talked almost entirely about politics, Geffen looked at me and asked, "So, do you like movies?" And while I do like movies, that it took two hours for us to broach the topic might have been my first clue that this probably wasn't the gig for me.

But it wasn't until a few days later—when someone who had worked with Katzenberg at Disney told me that he used to come in very early and feel the hoods of people's cars to see how long they had been there—that I knew I wasn't going to end up at Dreamworks, despite the amazing reputation of the principals and the glamour of the business. I had just left a job that owned me; I wasn't looking for another one. Instead, I created a package that included cohosting the political talk show *Equal Time*; working as the Washington editor of *Vanity Fair* magazine; and giving speeches around the country. It was fun, flexible, and lucrative, and I couldn't have been happier. I met a great guy, got married, and had two beautiful kids; I've never looked back.

That's my story. Certainly, I wanted a family. But it wasn't just that: My White House experience had left me feeling more than a little burned out, and I wanted a different kind of life. So I traded power for control, intensity for flexibility. Plenty of women have made different choices—or have made similar choices for different reasons. And just because some women don't want the big honkin' all-consuming eighty-hour-a-week job, that doesn't mean that none do. A lot of men don't want those kinds of jobs, either. And a lot of women not only want them, but are wildly successful at them. Take Meg Whitman, the CEO of eBay, or Michelle Bachelet, the president of Chile, or Oprah Winfrey. Or my friends who are members of Congress, or doctors, or lawyers at high-powered firms, or entrepreneurs, or actors. Assuming that women—even women with children—don't want the top jobs means that too many women will never get the chance to make those important decisions for themselves. Employers will make assumptions based on what they *think* women want, and calcified stereotypes will keep the doors stuck shut.

So women are rational. Different domestic demands and professional opportunities (and rewards) create different incentives for men and women. And as a result, they often make different choices. But what about interests? Is there any evidence that gender affects what captures our imaginations, touches our hearts, or shapes how we choose to spend our time?

To be sure, we all have our own particular interests, and they change and develop over time. When I was in high school, I did well in math and science. I had no idea what I wanted to be when I grew up, but my college—Santa Clara University—required all freshmen to declare a major. I'd had a wonderful biology teacher in high school, Mr. Stanford. And even though I got a D on my insect collection—a project that required me to catch, kill, pin, and label all kinds of creepy-crawly things in a large Styrofoam-bottomed box—I loved the class and chose biology as my major. But by the end of that first year, I realized I was infinitely more interested in how people organized their collective life than in the life cycles of worms, so I switched to political science. It never occurred to me that any of these decisions were related to my gender—and maybe they weren't. But extensive research shows that there is an "enormous average difference" between women and men in their desire to work with "people versus things." According to Steven Pinker, a psychology professor at Harvard, this difference causes people to make different career choices.

A University of Michigan study found that young women viewed "pure math and physics careers as isolating and not so helpful to society. Since they saw themselves as people-oriented, they chose biology research or health instead." And that may help explain why, even as there has been an enormous increase in the number of women receiving all levels of university degrees across the past thirty years, there has been little change in the percentages of women choosing particular fields. In 2003, women earned 65 percent of doctorates in education, 54 percent in the social sciences, 46 percent in biology, 18 percent in physics, and 17 percent in engineering—a pattern that is virtually unchanged in more than twenty years.

Moreover, girls who scored extremely high on the math portion of the SAT often also scored high on the verbal portion, a finding not necessarily true of mathematically gifted boys. And students with high verbal scores were less likely to pursue math or physical science; they just didn't think these fields would allow them to use all their talents. Because the truth is, much of math is now theoretical, and physics doesn't require as many "people" skills as, say, medicine. Which may mean fewer girls will be interested. Or it may mean we need to find new ways to make disciplines like physics and engineering more appealing to women's more diverse interests. Maybe if the field's most visible leaders talked about the practical, "people-oriented" benefits of the physical sciences—like how rural villages in Africa might get clean water and affordable energy—more girls would be interested.

The numbers make an interesting case. But they don't "prove" that men and women, by their natures, have different interests—and they don't rule out the effect of other factors, like stereotypes, social expectations, and opportunity. Forty years ago, it was widely believed that boys were vastly more interested in playing sports—and that that alone accounted for the huge imbalance in male and female athletes in the nation's high schools and colleges. Then along came Title IX. As opportunities for girls to play skyrocketed, so did their interest in doing so; over the next thirty-five years, the number of women playing high school sports increased a whopping 904 percent, and the number of girls playing at the college level rose 567 percent. Old, settled ideas and accepted truths had to be completely re-examined.

When I asked Shirley Tilghman, the president of Princeton and a molecular biologist by training, why there are more women in her field, she offered her own explanation. "I actually have a theory, and I can actually argue against it as well as for it, that one of the reasons molecular biology was such fertile ground for women is that it was a brand-new science, there were no norms about how we behaved in this science. If you think about when my generation got into molecular biology in the late 1960s and early '70s, which was really when the field was just taking off, there were lots of opportunities, and it was relatively easy to get a job."

Tilghman said she also found the field fascinating. "In chemistry, which is where I'd been, there was a sort of sense in my mind—and I don't want to disparage chemists—but it felt as though it really was a mature field, that the most interesting questions had been answered. Whereas in biology, the sky was the limit. It's the way neuroscience feels right now. If I was twenty-two years old right now, I'd be a neuroscientist. I think the culture of engineering, even by the 1960s, was really a male culture. Pocket protectors, and slide rules, and guys with glasses with tape. And even though those are superficial descriptions of the field, it was a field that probably would not have felt very welcoming to women. But biology was just too busy to worry about whether it was okay that as the field grew exponentially, increasingly it was women."

In other words, Tilghman went where she thought the questions were most interesting—and the opportunities were the greatest. But she quickly admits that computer science—a field that exploded at virtually the same time as molecular biology—"undermines my argument." Initially, women appeared to be going into the field, she explained. But over the years, the numbers have actually declined. "The only answer I have for that is, it's about what happens in the bedrooms of thirteen-year-olds. While girls at the age of thirteen are off doing all sorts of things, there are an awful lot of boys in their bedrooms, with the shades drawn, hacking into the *New York Times*."

From Assumptions to Options

Okay, the pipeline leaks; how do we fix it? To begin with, employers have to continue their search for innovative ways to make the workplace more flexible. New technology and a willingness to rethink job requirements have helped. But technology has also created new obstacles. In a world where information flies around the globe in real time, workers are expected to answer calls and e-mails 24/7 or to travel extensively, often on short notice. Still, committed employers can make a difference.

"I think there are a whole series of things you can do," says Tilghman, of making Princeton more family-friendly. "The first is to make it clear that you would fully expect faculty to have a family. The expectation is not that every woman on the faculty is going to be celibate, with monklike devotion to whatever her field is. And I think that symbolism should not be underestimated." Changing the culture is critical, but just the first step; Tilghman also put together a committee to find and implement concrete ideas. "I went years thinking there was a silver bullet. And the answer is: There's no silver bullet. There are these little things, and the key is to just keep finding them and solving them. They add up."

Now that I have two children of my own, I value my flexibility more than ever. Of course, I have deadlines to meet and appearances to make, and occasionally, trying to mesh those with the needs of my family gets hairy. But far more often, I can catch the school play or take a sick child to the doctor without jumping through hoops—or covering my tracks. And if I have to go back to work after the children are in bed, that's a small price to pay. Flexibility makes my life manageable.

Judith McHale, the former CEO of Discovery Communications, was a champion of flexibility during her tenure. Not only was it a benefit for her employees, regardless of whether or not they were parents; it also made the company more competitive. "It was absolutely a conscious decision," she told me. "Obviously my own perspective as a working mother would have impacted me . . . But it was actually a strategic, quite definitely a strategic imperative for the company—and I presented it to the board that way roughly ten, twelve years ago. It was the height of the Internet bubble, and we were competing with AOL and everybody else out the Dulles corridor [the area just outside Washington, DC, where a lot of technology companies are located].

"We are a privately held company. I did not have stock options to offer people, and those were hugely attractive. Nor did we have buckets of cash to throw at them. But the one thing that I felt we could offer them was time. Based on all the research, a high premium was placed on time, you know, free time, time to do other things." In short, flexibility became a way to make Discovery more competitive, she said. And telecommuting, flexible hours, and on-site health care became hallmarks of the company culture.

Senator Kay Bailey Hutchison became a working mom when she adopted two children in the 1990s. Her schedule is the opposite of flexible; when there's a vote on the Senate floor at two o'clock in the afternoon or two o'clock in the morning, she has to be there. It doesn't matter if her kids are sick or starring in the class musical. But not all workplaces need to be similarly rigid. "You can have flexible hours where someone can come in at six or seven in the morning and leave to be home when their child gets home from school," she tells me. "Those are the things that really are helpful. Having day care on site is a huge help for working mothers. And we need to continue to improve on that because every society that has women as equals in the workplace is more prosperous. Obviously, if you're using 100 percent of your brain power, you are getting ahead. We need to have the flexibility—and that's not to say that we shouldn't also have total support for women who want to stay home and raise children. Because it's so wonderful to have that, and I think that women should have the choices and we should make it easy—as easy as possible—for them to make the choices. Raising children is hard, but the better we can accommodate women either working or choosing to be full-time mothers is very important for our society."

A recent study of seventy-two large U.S. firms showed that family-friendly policies increased the number of women in senior management positions in subsequent years. But companies have to do more than put those policies in place; they have to encourage both women—and perhaps more importantly, men—to take advantage of them. Benefits perceived as "women only" can actually become career obstacles. Alexis Herman, the former secretary of labor, said that the men on her staff were the first ones to take advantage of policies like Family Medical Leave. "The men loved it," she told me. "And it wasn't a stigma. I made it okay."

A few years ago, the *Harvard Business Review* generated a lot of attention with a study called "Off Ramps and On Ramps: Keeping Talented Women on the Road to Success." The survey of mid-career women who held graduate degrees or college degrees with honors found that 37 percent had taken extended breaks from work, stepping off the fast track, on average, for a little over two years. Most of them wanted to return to work, but less than half could find full-time jobs.

"The old idea was, all you needed to do was fill the pipeline with women and wait around for a couple of decades for them to move through the ranks. [But] there's an enormous amount of leakage from the pipeline—once women off-ramp for even a short while, it's incredibly difficult to get back in," says Sylvia Ann Hewlett, lead author of the study and president of the Center for Work-Life Policy.

Leaving talented women stranded on the side of the road is a flat-earth approach to the changing global economy. It's shortsighted—and it's hypocritical. It penalizes women for doing what so many people tell them is most important: being good parents. Since the Harvard study appeared, some companies are starting to wake up and smell the opportunity, and many are finding new ways to maintain or restart relationships with women who choose to step off the fast track for awhile.

Business schools are also getting into the act, experimenting with on-ramping courses. The first time I opened *Newsweek* and saw an ad for a program called "Back in Business" at Dartmouth's Tuck School of Business, my heart did a little pitter-patter. Across a photo of a forty-something woman in a stylish business suit walking on a narrow beam (a confidence building exercise, perhaps?), the bold caption reads "Restart your career. Reinvent your future." Companies that want to participate have to do more than talk the talk; they have to redesign actual careers to fit the needs of women returning to the workforce.

Maybe, just maybe, companies are starting to realize that this is a place their interests and women's intersect. Says Eliza Shanley, cofounder of Women@Work Network: "There's a general sense among employers that whoever figures this out first wins."

Making work work for women isn't about creating another entitlement; it's about economic self-interest. In an increasingly competitive world, companies can no longer afford to write off half the talent pool—or to let qualified employees walk out the door. A few years ago, Domino's pizza found that replacing a single hourly employee cost them $2,500, and they launched a major initiative to reduce turnover. If holding onto the pizza delivery guy makes economic sense, doesn't it follow that holding onto promising young associates in law and accounting firms, talented managers, and high-powered sales staffs is a smart thing to do? Increasingly companies are finding that it is.

Underlying this effort to restructure the workplace is the notion that there isn't a one-size-fits-all way to accommodate women—or for women to mesh career and family. Some, like former Supreme Court Justice Sandra Day O'Connor, will want to have children when they are in their twenties, take time off when their children are young, and then start or restart their careers. Others, like Governor Kathleen Sebelius, will find interim ways to pursue their professional goals, while raising their kids. "My kids were little, and I was working for the trial lawyers, traveling a lot, working sixty hours a week and had these two little boys,"

Sebelius told me. "I lived in the capital city [Topeka], and the legislature in Kansas is a part-time position, ninety days, and I could come home every night. My boys were two and five when I ran, and I thought, 'Well, this actually would be a great part-time mom job.' This was my mommy track, literally, to step out of the high-paced, travel-intense world and into, really, a much more part-time position. So I ran in '86."

Other women, like me, will start careers right out of college, and keep the pedal to the metal for a decade or more. I was thirty-one when I went to work at the White House, thirty-five when I got married, and thirty-eight and forty-one when my children were born. Each of these options has advantages—and drawbacks. And women should be honest about what these are. Only 2.5 percent of babies in the United States are born to women over forty. Two and a half percent! And while new technology has helped reduce infertility among younger women, it hasn't been nearly as effective for older women—Madonna, Geena Davis, and Elizabeth Edwards notwithstanding. They were lucky, and so was I. I now counsel women to consider the risks, really consider them, if they think they can put off having children. That said, the choices I've made have worked out great for me (at least I think so on *most* days). I burned through a lot of ambition before I became a mother, which made it easier for me to let up for a while. And I'm far more patient than I would have been a decade earlier. I also had enough experience that I could create a career path that fits my needs at this particular stage in my life. And I fully expect those needs to change as my children get older and my interests evolve.

"Part of it is recognizing that fifty is probably the new thirty, and that workers—whatever they're doing—are not in the same situation that we were twenty-five years ago," Sebelius explains.

"Then, a lot of workers chose something at twenty-one or twenty-two. The goal was to work your way up as high as you possibly could, get your gold watch and pension at the end." But now, people will hold six or seven or eight jobs during the course of their lives. "Not just politicians, but everybody," she says.

"And so I think women are wonderfully positioned to say, 'As part of my six jobs in my life, one of them as mother, one of them as this, I'm going to take a stand in the city. I'm going to be mayor.' And if that leads to something else, great. If it doesn't, perfect."

Of course, some women decide not to have children at all. "I could not have done my job and raised children in the way that I would have wished to raise them," says Sherry Lansing, the former head of Paramount. "Now that doesn't mean I'm right. I'm just saying that *I* personally couldn't have done it. I believe you have to make choices in life, and I'm really content with the choices that I made. But I think when you try to do everything you feel pulled. Just something's got to give. So I always thought you could do two out of three, or you could do all of them sequentially. And when I used to give speeches, often to women's groups, when I used to say that I don't think you can have it all, I used to get '*Sssssss*.' Today when I say it I get a round of applause. And the word that people are using is 'sequential.'"

CHANGING THE PARADIGM

Women want and deserve not only the flexibility to manage work (and family) from day to day, but also the ability to make choices that allow them to pursue their goals across a lifetime. Whatever those goals are. However they choose to pursue them.

Some women may choose to stay home after their children are born, others to work part time or full time. Some may choose not to have children, others to pursue less demanding careers so they can devote more time to other interests, whether that means climbing mountains or driving the kids to soccer practice. And women need to accept—and yes, support—other women who make choices different from their own. The so-called Mommy Wars, where stay-at-home moms and working moms question each other's priorities across a chasm of mistrust, are as real as they are destructive to both sides' shared interests. Too often, they're rooted in a grass-is-greener insecurity that makes women wonder whether they're doing what's best for themselves and their families.

Like most of my friends, I've sometimes struggled with the choices I've made, alternatively casting an envious eye on the women who have chosen not to work—and then at those who have accomplished more than I have professionally. First, it's important to acknowledge that every choice involves sacrifices. Those of us who downsize our career objectives to spend more time with our children give up the stimulation, satisfaction, and remuneration that we often got from our successful, pre-mommy work. Those of us who trade family time for the joy (and often economic necessity) of work miss the important milestones and ordinary events of our children's lives. So at the end of the day, each of us needs to decide: How much time can I spend away from my children and still feel like the mother I need to be? An honest answer to that question is a critical tool in making other decisions.

"In 1991, I left my job as a correspondent for *60 Minutes* because I wanted to spend more time with my family," says Meredith Vieira of the *Today* show. "I think that decision disappointed a lot of women because I was basically saying, 'I can't have it all and I don't want it all.' I wanted to set some priorities in my life, so I went with what matters most to me, family. I never lost a night's sleep over it. But I remember one woman cornered me at a party and quite angrily said, 'You've set back the feminist movement.' My attitude was that I would have truly set back the movement had I lived a lie. You have to be true to yourself."

At the same time, women need to accept that there are many different ways to raise successful children. I have friends who never took their foot off their career accelerators, even when their children were young. One friend told me how she watched her own mother suffer from boredom and drift as she struggled to raise five children. "I'd come home from school, and she'd be sitting on the couch in her bathrobe, drinking coffee and watching TV. And I knew then that I did not want her life." So my friend has worked in a series of demanding and rewarding jobs, while raising two incredible sons. I have other friends who left high-powered and high-paying careers when their first, or second, or third child came along. Each has made her decision for her own reasons. And each of those choices has to be respected.

Yes, the problems persist—even after decades of debate about the causes and attempts to find the cures. And progress that had once seemed inevitable now seems stalled; women still find their path to genuine equality littered with obstacles real and perceived, cultural, and perhaps innate. Still, I find myself optimistic about the future.

At a recent campaign stop, Hillary Clinton said she'd been touched by many of the elderly women who had approached her at various events. "I remember the woman who said to me, 'I'm ninety-five years old,' as I was shaking her hand. She said, 'I was born before women could vote, and I'm going to live long enough to see a woman in the White House.'" The trajectory, indeed the sweep of history, seems to be inevitably and unstoppably in the same direction.

And yet, it hasn't been easy. At forty-six, I'm not young, but maybe I'm just young enough to have avoided making the sacrifices—from mastering the nuances of zone defense, to hiding (or not having) my children—that so many women, some just a few years older than myself, made. I may have also avoided their outsize expectations—and their outsize disappointments. Change still isn't easy. And it still isn't coming as fast as many of us would like. But I'm confident it will keep coming. Not just because it will make for a happier workplace, but because it will make for a more productive one.

Which brings us to the question: How much change is enough? What, exactly, is our objective, and how will we know when it has been achieved? For starters, I think we have to abandon the idea that even if all the obstacles could be eliminated, there would be an equal number of men and women in virtually every profession. We don't need gender parity among elementary school teachers or bond traders before we can declare victory. We don't even need it among physicists. That's not to say that it won't happen. It could. But isn't it more likely that even if we eliminate the conflict between having a high-powered job and having a family, unravel the mysteries of innate aptitude and interest, and root out discrimination, there will *still* be more women in social psychology and more men in engineering? And isn't that okay? I think it is.

Chapter 8

Closing the Confidence Gap

"Nobody can make you feel inferior without your permission."

—Eleanor Roosevelt

April 19, 1993. Every television in the West Wing was tuned to the same scene: flames shooting from the roofs and windows of a cluster of buildings somewhere outside Waco, Texas. As I stared at the set in my office, horrified, I was besieged by calls from reporters. Would the president be making a statement, they wanted to know? No decision, I told them. But I strongly believed he should.

Earlier that day, federal agents had stormed the compound of the Branch Davidian cult, ending a standoff that had lasted more than two months, cost the lives of four federal agents and wounded sixteen others. The time had come to take action against David Koresh, the cult's leader, who was using the more than eighty people inside as human shields. Amid rumors that the children were being sexually abused and concerns that the long standoff was draining FBI resources, Attorney General Janet Reno had told the president a day earlier that the FBI wanted to raid the compound, and he had given her the green light. But things had gone terribly wrong. And when it was finally over, everyone inside was dead.

Though the decision was technically the attorney general's, the president was her boss, the leader of the federal government, the highest official in the land. What's more, Reno had sought—and been given—the president's okay. I thought he needed to be accountable—even if he wasn't completely responsible. Anything less and he would be accused of passing the buck. I tried hard to convince him, and the senior staff, of my view. I wasn't alone; Bruce Lindsey, a senior aide to the president, and several others agreed with me. And for a while, it looked as if the president did, too. But eventually, he decided to let Reno go out and face the cameras, while the White House did nothing more than issue a statement.

That turned out to be the wrong decision, and even though I argued against it, I felt that it was my fault. What good was I if I couldn't persuade people—the president included—about something that had seemed so obvious to me? I certainly didn't expect the president to side with me every time; no one's judgment is that good.

But why didn't he listen? Was it because I was a woman? Was I too young or too uncertain of my own opinions? Did I speak too softly—or not softly enough? Could I have said something different to him that day that would have convinced him to face the cameras?

The questions gnawed at me. Years later, I came across a passage about the incident in President Clinton's memoir, *My Life*. He recounted that I had urged him to talk to the press and take responsibility, and he regretted not doing so. I was surprised—and gratified—to learn that even though he hadn't taken my advice, the president had remembered it. But it wouldn't change the way I felt at the time: The whole episode made me question my own effectiveness—and it sanded another layer off my confidence.

I often felt that I had to work harder than my male colleagues to be heard, a frustration I know so many other women have experienced. Why? In my case, it didn't help that I was the first female and one of the youngest White House press secretaries in history and that I was from California. I often call that the trifecta of how *not* to go to Washington. It was hard for me to establish my bona fides—especially when I was trying, like so many women, to navigate that elusive line between being authoritative and being a bitch. Too often, I erred on the side of trying to be liked.

During my briefings, I developed the bad habit of saying "I think" before answering a reporter's question. As in, "I think we'd like to work with the Congressional leadership from both parties to pass a comprehensive heath-care reform plan." It wasn't that I didn't know that statement was true; rather, I was trying to soften the exchange. To me, it felt like the verbal equivalent of a lob rather than an overhead smash, and I wrongly thought it might ease some of the tension in the briefing room. I finally realized it was having the opposite effect when Carl Leubsdorf, a reporter with the *Dallas Morning News*, sputtered at me one day, "You think? You think? Why don't you know?"

Throwing Like a Girl

What happened to me in the briefing room—and sometimes in meetings with the senior staff, the cabinet, or the president—was in a way basic and biological: I threw like a girl because I am a girl. The truth is: Women do certain things differently. That's not to say all women are one way, and all men another. Of course not. But on balance, women tend to communicate a little differently. We're usually less willing to blow our own horns or draw attention to our accomplishments. We're less inclined to take certain risks, but more likely to share credit. We're more practical. There are complicated reasons for this, some bequeathed by Mother Nature, others carefully taught. Regardless, the effect is that women can seem less confident, less sure of themselves. And that, in turn, can make it harder for them in a world that is still largely shaped by men. A world where "I think" can sound weak, rather than generous.

If a male press secretary had developed the same bad habit, I honestly believe he would have been treated equally harshly. But I also believe a man would have been less likely to go there.

Lasting patterns of communication are often established in childhood and driven—at least in part—by biology. According to Dr. Louann Brizendine, girls are more comfortable making decisions together, and they often use an "affiliative" style of speech, the kind of language I hear my daughter and her friends use all the time. "Let's go upstairs and play with our American Girl dolls, okay?" one of them will say. It's part suggestion, part question; it seeks consensus before action.

"[Girls'] genes and hormones have created a reality in their brains that tells them social connection is at the core of their being," Brizendine explains. Boys know how to use this affiliative speech style, too, but research shows they usually don't. "Instead, they'll generally use language to command others, get things done, brag, threaten, ignore a partner's suggestion, and override each other's attempts to speak." Boys are "not concerned about risk or conflict. Competition is part of their makeup. And they routinely ignore comments or commands given by girls."

Clearly, girls and women understand how to use the command style of speech, too: Just head to the German Bundestag, where Angela Merkel serves as the first female chancellor, or to the nearest playground, where the moms routinely take charge of their kids. But in other cases, they're less likely to. And it creates something of a confidence gap.

Rosabeth Moss Kanter, a professor at the Harvard Business School, was asked once whether she thought the men in her classes were more confident than the women. "The women are very bright, very motivated," she said. "There is no talent difference that I can see. There is sometimes a difference between the men and women in the willingness to claim airtime in class. The men seem to feel that they can start talking and eventually they'll have a point to make. The women are more likely to feel that they ought to have something valuable to say before they say it."

That pattern seems to hold true in almost any setting where there is a mix of men and women. "Confidence is the expectation of a positive outcome," Kanter continues. "If you think you're going to be criticized and attacked if you don't have all the facts, you hold back. If you hold back, you don't try, and you don't know how good your ideas are."

"Middle school is the moment of bifurcation," says Fern Marx, a senior research scientist at the Wellesley Centers for Women.

"Girls and boys begin to differentiate academically and in terms of interest." They also begin to differ in terms of confidence. "A girl without the right answer believes she's dumb. A boy just thinks he's unprepared."

Believe it or not, you can ratchet down women's scores on math tests just by reminding them that they are women. The phenomenon, called "stereotype threat," was first identified by psychologists Claude Steele and Joshua Aronson at Stanford University in 1995. They found that if they asked African American students to identify their race before taking a standardized test, they only got half as many questions right. *Half.* Since then, dozens of other studies have confirmed that getting people to think subconsciously about their race or their sex can activate negative stereotypes—such as those that suggest African Americans don't do well in school or that women aren't good at math or science—and cause them to perform poorly.

But is the opposite also true? Can you override stereotypes by getting people to think about their strengths? The answer to that question also appears to be yes. A few years ago, psychologists took a group of ninety college students—half men and half women—and divided them into three groups. Before giving them a test, they asked each group to answer a different set of questions. The first was asked whether they lived in single sex or coed dorms, which subtly cued them to think about gender. The second was asked to write about why they chose a private liberal arts college, which primed them to think about how smart and accomplished they were. "We were activating their snob schema," Matthew McGlone, one of the researchers, said. And the third, or control group, was asked to write about a more benign topic, their experience living in the northeastern United States. All the students then took the Vandenberg Mental Rotation Test, which measures spatial ability—and where men perform significantly better than women.

When researchers analyzed the results, they found that the men in the control group did 15 to 20 percent better than the women, which was consistent with previous findings. Of those in the group that had been primed to think about their gender, men outperformed women by an even bigger margin, scoring 25 to 30 percent higher. And most surprisingly, among those who were cued to think of themselves as students at an elite private college, there was no difference; women's scores improved substantially while men's stayed the same. In other words, by triggering women to think positively rather than negatively about their accomplishments, the gender gap was slammed shut.

Which begs the question: Why did encouraging students to preen a bit improve women's scores, but not men's? Somehow, the question reminds me of a cartoon someone sent me a couple years ago that showed a woman and man, each looking in the mirror. The woman was curvy and cute, but in her reflection, she saw a person with a backside the size of a small blimp. Meanwhile, the man in the cartoon was downright fat (and balding). But when he looked in the mirror, he saw an Adonis looking back. The caption read: "The Difference Between Women & Men."

Good Girls Don't

So it starts early—and holds fast. Too many women don't raise their hands in class, don't volunteer for new assignments, don't seek deserved raises and promotions.

"Women negotiate very effectively on behalf of their companies, but not on their own behalf," says Victoria Medvec of the

Center for Executive Women at Northwestern University's Kellogg School of Management. "A common mistake women make is not asking for the tools they need to be successful, like staff increases and other resources," she says. As a result, their efforts to succeed and advance are undermined.

According to Linda Babcock and Sara Laschever, authors of *Women Don't Ask: Negotiation and the Gender Divide*, men think of negotiating like a ballgame or wrestling match, while women think of it as a "trip to the dentist." As a result, men initiate negotiations about four times as often as women. And the consequences can be staggering: One study calculated that women who consistently negotiate their salary increases earn at least $1 million more during their careers than women who don't. What's more, while women own roughly 40 percent of small businesses in the United States, they receive just 2.3 percent of the equity capital; the other 97.7 percent goes to companies owned by men. Why? Because women don't ask.

Women need to get over it; they need to get a handle on what they're worth and what the market will bear. And they need to be clear about what they want. But they can't always ask the same way men do. "I think women need to find their own negotiating voices," Babcock explains. "Our society still has a double standard for the behavior we accept from women and behavior we accept from men. So take an example where a woman gets another offer. She's at her current job and she comes into her boss's office, and she says, 'If you don't match my salary, I'm out of here.' That approach, which a man might be able to get away with, may not fly with a woman. So she needs an approach which would be something like, 'Hey, I got this other offer. But I like working here, and I'd really like to stay. Can you find a way to match the offer?' So it's a little bit of a softer approach because people don't accept a really aggressive approach from a woman."

Women are also too reluctant to recognize and take credit for their accomplishments. In politics, men are much more likely to "self-identify," that is, to run for office without being asked. If you lay out the five or six or seven criteria required to be a successful candidate, women will look at the list, point to the one qualification they *don't* have, and decide not to run. Men will look at the same list, find the two or three things they *do* have, and assume they're going to run—and win. And it's true whether they're running for the first time, or considering a higher office after they've already been elected.

"Women tend to run because they're concerned about an issue; they don't wake up thinking they want to be governor the way men do," says Jeanne Shaheen, a former three-term governor of New Hampshire who is now running for the U.S. Senate.

Studies show that women are even less likely to run if the race looks competitive or if the other candidate is a man—even though when women do decide to run, they win. Research from Jennifer Lawless, a political scientist at Brown University, shows that women are just as likely to win elections as men. Moreover, even though women make up more than 50 percent of voters, they still represent only a quarter of elected officials nationwide. Equally likely to win. Half as likely to run.

Of course, there are exceptions. Kathleen Sebelius was a Kansas state legislator when she decided to run for statewide office in 1994. After ruling out running for treasurer (the job had been stripped of much of its power), attorney general (she wasn't a lawyer), and governor (she wasn't ready), she decided to run for insurance commissioner. "I knew how we could win; and it was pretty simple," Sebelius told me. "You had to fire this guy (the incumbent) and at least credential me enough that I was a legitimate alternative. And there were plenty of reasons to fire this guy." So she ran and won.

In hindsight, it looks like an easy decision, but at the time, it was loaded with risk. Sebelius is a Democrat, and Kansas is a Republican stronghold. In the presidential race two years earlier, president George H. W. Bush crushed Bill Clinton in the state. And while Democrats were occasionally elected to statewide office, no Democrat, let alone a woman, had *ever* been elected as insurance commissioner. Never. But after looking at the race, and her future goals, Sebelius decided to do it.

"I knew at that point that someday I might want to be governor," she explained. "My dad had been governor [of Ohio]. He always described it as the best job in the world, much better than Congress . . . And I just thought the opportunity to maybe do that someday would be great. But I knew I needed some credentials I didn't have. I'd done a lot of work on children's issues and health issues in the legislature, but I didn't really have any financial credentials. And in my own mind, I knew that the business boys would never like a woman unless you could kind of come credentialed. So the insurance commissioner was an office where I thought I could do a lot on health care, which was just starting to explode in 1994, and also have this sort of financial services credential, which might be very helpful in the future.

"I just made it clear fairly early on that I was running. If anybody else wanted to run, that's fine, but this was something that I was going to do," she says.

Unfortunately, the choices Sebelius made are too often the exception rather than the rule. "There aren't enough women who say, 'Boy, do I know a lot,'" says Swanee Hunt. "'I know how to organize a family. Not only that, I know how to organize a church.

Not only that, I know how to organize a drive across the community. Let me at the public policy!' But they don't do that."

Rethinking Risk

In short, men and women tend to see the same qualifications differently; in that great poker game called life, they see different possibilities when they're looking at the exact same cards. Ask virtually any woman in the Capitol and she'll tell you the same thing: As soon as a man is elected to the U.S. Senate, he starts hearing "Hail to the Chief" playing in his ear. It doesn't matter where he came from, or how much political experience he has; he automatically assumes he's qualified to be president. The women? Not so much.

Margaret Chase Smith, the first woman elected to the U.S. Senate, was once asked what she would do if she woke up in the White House. "I'd go straight to Mrs. Truman and apologize. Then I'd go home," she said.

Not John Edwards. The former U.S. senator from North Carolina was first elected in 1998; he was just forty-four years old, and he'd never run for Congress, the state legislature, or even the school board. He'd spent the previous twenty years working as a trial lawyer, where he'd honed his speaking and debating skills—and built his personal fortune. But he hadn't been very active in politics. Still, he ran—and he won. And a little more than four years later, with less than one full term in the Senate under his belt, he decided to run for president. He almost won the Democratic nomination. And when John Kerry chose him as his running mate, he was almost elected vice president. Four years later, he decided to run again.

Now I like John Edwards. And I like his wife, Elizabeth. And for several years, his two youngest children were playmates of my children. John's a good man, and I'm not criticizing him. In fact, I admire his boldness, and I wish that more women were willing to take similar risks, to decide that they're the best man for the job, and to raise their hands for campaigns, promotions, more responsibility, more money. After all, a lot of times, half the battle is getting into the battle. And that's not only true in politics, but in virtually every facet of public life.

Judith McHale, the former CEO of Discovery Communications, originally joined the company as a lawyer. But soon, she went to her boss, founder John Hendricks, and asked for more responsibility. "I said to him, 'Don't just think of me as a lawyer. I'd like to do other things as well.' And then he immediately put me in charge of HR, Admin, and IT, which was so hysterical. But that was good. And then I slowly but surely learned the whole company because he and I worked so closely together."

Too few women do that, she says, even in a company like Discovery where the CEO, COO, and more than half the senior managers were women. "We post every position in the company ten days before it's public, so everyone has a shot at it and is encouraged to apply. Four or five years ago, we had an opening for a major programming position, and no women applied. And I thought, 'God that is so weird in this company.' And so our head of HR and I started looking at it, and what came out pretty universally was this feeling among women, 'I thought if you thought I would be good enough for the job you would have asked me.' That was a really scary moment."

Of course, not all risks are created equal. And if men are more likely to take the kinds of risks that create professional opportunity, it's also true that they're vastly more likely to take the kinds of risks that leave women (and a lot of men) dumbfounded. The Darwin Awards—named for the evolutionist—are given every year to those who "improve the species . . . by accidentally removing themselves from it." And many of the recipients' feats are nothing short of spectacular. One man, after a night of drinking with friends, bet that a detonator he found wouldn't work. He put it in his mouth and asked friends to trigger it. They did (a pretty amazing detail in and of itself), and the thing blew. Adios, amigo. In another incident, two guys wanted to see who could hang longer from a freeway overpass. Unfortunately, the winner was too tired to pull himself up, and he fell onto the hood of a passing semi truck, bounced off, and was run over. He, too, was sent to his eternal reward. Predictably, virtually all of the Darwin Award winners are men.

Not all bad risks will get you killed, of course. But some *will* cost you money. And interestingly, women may be less willing to take those kinds of risks, too. The pervasive stereotype is that men are better at all things financial than women. And that attitude starts early. A few years ago, one of the major financial houses did a survey of high school students, asking them how good they thought they were at math and money. The boys concluded they were "pretty good" and the girls "not very good"—even though their actual knowledge was exactly the same! Certainly, Wall Street is a male bastion, and only a handful of women have ascended to the storied ranks of top dealmakers, money managers, and investors. But what about the less-rarified ranks of ordinary investors? Numerous studies show that women actually achieve better returns than men (see "The Smart Thing" in chapter 2).

Despite the sometimes discouraging numbers, Alexis Herman, the first African American to serve as secretary of labor, believes women *are* great risk takers. But, she says, "I think we as women engage in a lot of what I call 'protective hesitation.' The average woman—to be sitting where she is today in a leadership position—took a lot of risks to get there. They had to work through a lot of situations just to be where they are. So we don't even have the appreciation for what we've done, do you see? I think we are great risk takers. We just don't see it."

Herman learned to take risks as a young African American girl growing up in the segregated South. Because her father had played professional baseball in the Negro Leagues—and had once trained Satchel Paige—he could attend games at Hartwell Field, where the AA Mobile Bears played. He couldn't sit in the all-white bleachers, but he and his family were invited to watch from the dugout. To get there, they had to pass through the players' tunnel that ran between the stands.

"The goal was to get past there, because people threw Coke bottles at us, and milkshake cups. I mean, it was like whatever they had in their hands," she told me during a conversation in her elegant Washington, DC–area home. "You knew they were going to attack you just for being inside the stadium. But the goal was to get all the way to the bench. And I would go. And my father would hold my hand tight. So I think for me, I can look back now and see I sort of took all of that as a part of what you had to endure if you were really going to stand up and fight for change and what you believed in. I just saw my father take so much growing up. And I think some of that he imparted to me."

So sometimes just showing up requires taking a risk, a chance, a leap of faith—especially for the first ones through the door, when there are no natural support networks, groups of friends or allies, or even anyone who looks like you. "Sometimes we as women don't realize we need it," Herman tells me. "When you function as door-openers, you don't know, sometimes, how hard that is on you."

As I think back on my own experience, especially my White House years, I know she is right. "I didn't have much time to dwell on it then," I tell her. "I was just scrambling to get through the day."

"It's like walking through those bleachers and having them throw Coca-Cola bottles on me. It's kind of what you had to do just to sit on the bench," she says with a chuckle. "I didn't think, 'So why are you getting hit on the head with the Coca-Cola bottles?' Ouch. And you just keep walking. Women don't know what it means to have that kind of support."

CREDIT AND CREDIBILITY

It doesn't help that women get less credit for the same accomplishments. The late Ann Richards, the truth-talking former governor of Texas, was fond of saying, "Ginger Rogers did everything Fred Astaire did, only backwards and in high heels." And yet.

The credit gap shows up in all kinds of ways, in virtually every field. It's certainly a fact of life in Washington. For example, there's a mentioning game that goes on before a presidential campaign, where pundits like me talk about who might get into the race—or who should get into the race. Hypothetically, a U.S. senator from a huge and important state like Texas, who had been reelected twice with more than 60 percent of the vote, who had previously been successful in business, had served in the state legislature, and had been elected state treasurer, would be on everybody's list of likely, even ideal, candidates. That candidate exists. Her name is Kay Bailey Hutchison. But I never heard anyone—myself included—say she'd make a great candidate for president in the run-up to the 2008 race. It's hard to imagine that would have happened to a Texas senator named Kurt Bailey Hutchison.

Don't get me wrong: I'm not totally faulting men for these lopsided evaluations. Men contribute to it. But so do women (and girls), who often downplay their own accomplishments and contribute to the sense that they're worth less. I was discussing this phenomenon with Shirley Tilghman, who says she sees it all too often, even among the most talented students at Princeton. A few years ago, she said, the campus paper carried a story about two incoming freshmen who had both been awarded the prestigious Westinghouse Science Prize. "The guy is interviewed," Tilghman said, "and he says, 'Oh yeah, I had this great project, and it was really exciting, and the judges loved it. It was so much fun.' His assumption was, 'Of course I won the prize.' And the woman was like, 'I was so surprised. My project, I didn't think it was that great.' Here are these two equally talented young scientists, and one believes he deserves to be there and the other one is just totally surprised."

The credit gap is also alive and well on the home front. The legal scholar Deborah Rhode contends that fathers get more credit than mothers for the same level of involvement with their children, which often works against women in child custody cases. And who hasn't seen it in the day-to-day expectations for both parents? A man takes his kids to the playground and then to lunch at McDonald's, and everyone marvels at what a good dad he is. But recently, a friend of mine moved with her husband and children from Washington, DC, to London, when her husband got a big promotion. They all arrived ahead of their furniture, which was making its way across the Atlantic so slowly it might as well have been in a canoe. For ten days, she entertained her kids, who had serial bouts of the flu, in a house with nothing but a couple of blow-up mattresses and a folding table borrowed from the new neighbors, while her husband worked fourteen-hour days. She took it in stride, just another day in the life. But if she had been a man, much fuss would have been made. I'm reminded of a routine that the comedian Chris Rock used to do, where he would describe men's efforts to get extra credit for the routine feeding, bathing, and caring for their kids. "What do you want, a cookie?" he would ask. "That's what you're *supposed* to do!"

This tendency on the part of women (with an assist from men) to underplay their talents and qualifications follows them throughout their lives. "You see all these boys who get C's in math and say, 'I'm going to be an engineer!' And all these girls who get A's in math and say, 'I'm not good enough,'" says Sally Ride, the nation's first woman astronaut. In an effort to convince girls that science isn't just a "guy thing," Ride created Sally Ride Science, a company that organizes festivals, summer camps, newsletters, and career guides. Clearly, she's a wonderful role model and an inspiration; she certainly was to me. Still, in 1983, when Ride rode the *Challenger* into space—and instant celebrity—she refused to let anyone write a book about her life; she just didn't think she'd accomplished enough.

Headhunters say that women often come off as less confident because they are more honest about their weaknesses, while men talk only about their strengths. "When it comes to professional modesty, women overdo it," says Marge Magner, for a time the highest-ranking woman at Citigroup. She said that when she would interview candidates for stretch assignments, women would often tell her they were not ready; men almost never did. "One of the things I tell women is, 'Listen, next time someone offers you a job, don't tell them you're not capable. Keep it to yourself.'"

Too often, women undervalue not just their own competence, but the competence of other women. Alison Estabrook, director of the Comprehensive Breast Center at St. Luke's–Roosevelt Hospital, said she didn't exactly hear "You go, Girl!" from all the women in her life. "There weren't a lot of female breast surgeons when I started out, and when I told a friend of my mother's that I was going to become a surgeon, she said, 'I would never go to a female doctor.' I didn't react. You have to have self-confidence. You have to know where you want to go. Women take criticism more to heart than men, and I think that can be a problem."

If women are going to get credit, they're going to have to take it. Sharon Allen, chairman of the board, Deloitte & Touche USA, counsels women to take responsibility for their own careers. "Don't assume that others are aware of your good work," she says. "When I was a young accountant, I was unhappy about not getting a promotion. I went to my supervisor and told him all these things I thought I should be given credit for, and he said, 'Well, gee, I didn't know you'd done all those things.' It was a real wake-up call. You don't have to be a bragger, but I think it's very important to make people aware of your accomplishments. I think women are sometimes less willing to do that."

Women may not get as much credit, but what they get, they seem more willing to share. Again, this may be part biology, part of women's hormone-driven need to establish connections with other people, to build and preserve relationships above all. Men, on the other hand, are more interested in figuring out the pecking order—and improving their place in it.

When the United States Tennis Association named its complex in Forest Hills, New York, for the legendary Billie Jean King, she was asked what her erstwhile nemesis Bobby Riggs might think. "Bobby is going to say he's responsible, which is fine. He did make a big difference in my life." In other words, she assumed Bobby would try to take credit, and she was happy to share it.

Debra Lee of BET remembers attending her first board meeting after being named chief operating officer of the company. "I did my reports," she says. "And Bob [Johnson, founder and CEO of BET] said to me afterward, 'You should say "I" a lot more. You always say "we." You should say "I."' But you know that is so counter—I mean, they know it's me. I want to give props to my team."

One of Hollywood's first successful female producers told me she believes that women's willingness to share credit was a key factor in their initial success. "A lot of the guys that were running it—this is so terrible what I'm going to say—they knew we really worked hard," she said. "They knew we were really smart, and they knew we really did great work, and we weren't threatening. This is a terrible thing that I'm about to say, and so they took credit. I don't think they really thought we were going to take over. And even when we took over, if I really asked, I think a lot of the guys that supported me taking over thought I would be a soft touch.

"I watched how we tended not to get our egos as involved and tended to find a way to make the deal rather than let, as I say, the dick get in the way. There's something that happens where these guys have to best one another and it gets in the way, and we tend to just want to get the job done. And I have noticed that difference. Yeah, and we just tend to concentrate on the work and not concentrate on who won. Now what I don't know is: Will that change when there are women all over the place? Because don't forget, we're mostly competing with men. And that's what I can't figure out."

BACKGROUND NOISE

So many of the women I've talked to—particularly those in male-dominated fields—have stories about being discouraged, sometimes subtly, often overtly. Shirley Tilghman told me an all-too-typical tale involving her first-year physics professor. "When I made a mistake in the lab—I made a mistake—and he said, 'That's why there are no women physicists.'" It's hard to imagine a professor being quite so blunt in today's more politically correct world. But the messages are sent and received just the same. "I suspect—I can't prove this, but I suspect—that very successful women in science figure out how to tune out the more subtle versions of these messages," Tilghman says. "They just don't see it. And it's part, I think, a function of self-confidence. You have to have some self-confidence in order to do that. It's also built on determination, but if you talk to a lot of women scientists over many, many years, they tend to report pretty much what I report, which is: It wasn't that these things weren't happening, but we chose to ignore them."

But it isn't always easy.

For most of my teenage years, I assumed the playing field was level—that girls were as likely to succeed as boys, both in school and beyond. I tried to dismiss evidence to the contrary as misinformed or worse—though the seeds, once planted, often grew deep roots.

One of my better—if at times opaque—high school teachers was Mr. Dickenson, who taught a year-long course called Humanities. Though an early bout with polio had left him with somewhat limited mobility, he could never resist an opportunity for a theatrical retelling of a good tale, and at his best managed to keep a class of sixteen- and seventeen-year-olds riveted. Once, I remember him galloping around the classroom, gesturing wildly with a six-foot wooden lance, as he relived the story of Sir Gawain and the Green Knight.

Another time, in a discussion of Renaissance art, Mr. Dickenson strode from the middle of the classroom, where he often lectured, to the front, where he drew two figures on the blackboard: an upside-down triangle—the male form—and a pentagon—the female form. "This is beautiful," he pronounced, tapping the triangle with the tip of his chalk. "And this is not." End of discussion. Mr. Dickenson never said "I think" or "some people think"; to him, the truth was as plain as the lines on the board.

I've always assumed that at least part of this lesson was an act; he clearly enjoyed provoking students, trying to get them to defend their own reactions and ideas. But I still wondered: Was this a Renaissance idea he was simply presenting to us? Or did he believe it? And was it, in the end, true?

In so many ways, and for so many reasons, women have been told that they're not as good, that their accomplishments aren't as valuable, that what's important to them is, well, less important. And too often, women have been willing to believe it. As a result, they're less eager to speak out, less willing to put themselves forward—and less sure of themselves when they do. And that creates a downward spiral, in which too many women never test the proposition; they never find out how good they really are.

It's up to women to change the dynamic. Women have to ask—for more money, more resources, and better opportunities. Not in the same ways that men ask, necessarily. Rather, women have to write a new script, one that allows them to get what they need, in ways that are consistent with their values and in keeping with their style. Women also have to be willing to own their accomplishments—and to talk about them. It's not an exercise in chest beating. It's a way to make sure that goals are met, best practices recognized, and credit is fairly distributed. In the same way a proud mother wants to see her child recognized for his or her achievements, she should want the same for herself—and for other women. Only when women take (and give each other) credit will their contributions be valued alongside men's. Only then will they close the confidence gap. Only then will they be able to really give the best, true measure of themselves.

This isn't what I think. It's what I know.

Chapter 9

Seeing Is Believing

"Men are taught to apologize for their weaknesses, women for their strengths."

—Lois Wyse, author of more than sixty books on marriage, motherhood, cooking, friendship, and business

When my daughter, Kate, was in kindergarten, she declared, "Girls can't be president; only boys." I asked her why she thought that, and she cited a book we often read called *If the Walls Could Talk: Family Life at the White House.* In it, there are anecdotes and historical facts about each of the forty-one presidents who have lived in the Executive Mansion, and of course, all of them are men. "But girls can be presidents' wives," she added, as though that were some consolation.

I explained to her that while no woman had been president yet, women were allowed to run for and hold our nation's highest office—and that someday soon, one would be elected. We talked about how silly laws had once kept women from doing lots of things, like voting or owning property. My seven-year-old greeted the notion that such laws ever existed with the exasperated eye-roll usually reserved for her little brother. And, of course, we talked about Senator Hillary Clinton, a name familiar to her from my past and our endless household conversations about politics. I hadn't endorsed a candidate, nor did I express a preference for any of the Democrats then contemplating running for president in 2008, though I'm sure I said I was proud of Hillary's efforts and believed that the nation was overdue for a truly serious female candidate.

Months later, a friend of the family's was over and asked Kate if she had a favorite in the presidential race then starting to take shape. Without missing a beat, she answered, "Hillary Clinton because she's a woman, and John Edwards, because of Jack and Emma." I was surprised—and more than a little amused. I suppose it's not a stretch that she would choose Hillary Clinton on the basis of gender. After all, her mother was writing a book called *Why Women Should Rule the World*, and my own prejudices on the relative strengths and weaknesses of men and women get plenty of airtime around the dinner table. But I'd never said I was for Hillary for any reason, including gender; still, she'd decided that having a woman as president would be a good thing. And why not? At the moment, her world is one in which women are as likely to be the boss as men. Her teachers and many of the administrators at her elementary school are women. Her dentist is a woman. Most of her friends' mothers work, as well as their fathers. She sees Katie Couric anchoring the evening news. Even several of the priests at our Episcopal church, as well as the presiding bishop of the entire denomination, are women. Clearly, her earliest experiences—and the expectations that will grow out of them—will be very different from mine.

Before most people can imagine themselves in a particular role, they need to see other people who look like them doing something similar. To be sure, there are exceptions; some people ignore the obstacles, the certainty of history, the voices that tell them never, and crash through barriers to create a new reality for themselves and those who follow. But for the most part, seeing is believing.

It was for me. I never would have become White House press secretary without the example, help, and encouragement of the women around me. Some of them I knew; others I only wanted to know.

I was just starting seventh grade when Billie Jean King accepted the challenge from Bobby Riggs, the self-styled Male Chauvinist Pig, for a tennis match that would be billed as nothing less than the "Battle of the Sexes." King, just shy of her thirtieth birthday, was at the top of her game—and she was leveraging her high profile to advocate for women in sports. After winning the U.S. Open in 1972, for instance, she said she would boycott the tournament the following year unless the women's prize money matched the men's. It did, and she played. (Sadly, it would take another thirty-four years before the men's and women's prize money at Wimbledon, arguably the world's most prestigious tournament, would be equal.)

I clearly remember the buildup to that match, as people chose sides based almost exclusively on gender, and Bobby Riggs seemed perpetually surrounded by live swine. Even at twelve, I felt the pressure. This was about more than bragging rights—and it didn't matter that Riggs was in his fifties and well past his prime. King said later that she felt that if she lost, it would set women back fifty years. "It would ruin the women's tour and affect all women's self-esteem," she said. But she didn't lose. She won. In a sold-out Houston Astrodome. In front of a live television audience of more than 50 million people worldwide. In straight damn sets. I still remember the thrill, the sense of pride, the feeling that something important had changed. It wasn't about tennis, certainly not to me. It was about forcing the world to take women seriously. To this day, every time I hear Billie Jean's name, I practically break into a chorus of "I Am Woman"!

A few years later, when I was a senior in high school, Margaret Thatcher became prime minister of Great Britain. At the time, my own political views were still evolving, but I remember the sense of possibility I felt as she stepped confidently onto the world stage. Then, the week after I graduated from college, Sally Ride became the first American woman in space, another episode I remember for the drama that surrounded it—and the satisfaction it brought. Each of these achievements broadened the range of possibilities for me, even though I knew I would never be a professional tennis player, a British prime minister, or an astronaut. These women, these pathbreakers, mattered to me.

Closer to Home

But it was the women closer to home who had the greatest effect on my expectations. A few months out of college, I decided I wanted to work on Walter Mondale's campaign for president. By now, my political views had solidified: I was a raging Democrat, and I'd had enough of Ronald Reagan's presidency. I didn't know anyone who worked in politics. I didn't even know anyone who *used to* work in politics. But I decided this was something I was going to do. So I called directory assistance and got the number of the California Democratic Party's Southern California headquarters; they then referred me to a woman named Terri Hanagan, who was the Mondale contact person.

It turned out she worked at a law firm where one of the partners, Mickey Kantor, was the state chairman of Mondale's campaign. We talked on the phone for a while, then Terri asked me to come and see her. I had to cancel our first scheduled meeting when my dog, a miniature poodle, was attacked by a German shepherd and I had to take her to the vet to get stitched up. As Terri loved to recount, this dog-ate-my-dog episode nearly derailed my political career. Fortunately for me, Terri gave me a second chance—and then an unpaid internship in her office, answering the phones, inviting dignitaries from the city's political establishment to greet Vice President Mondale at the airport, and other critical tasks. It was grunt work, but I loved it. To pay the bills, I got a job in a department store, selling china evenings and weekends. But I spent as much time as I could working on the campaign, and Terri patiently coached me, introduced me to people around the office, and even invited me to her house for dinner. She was my anchor in those early days, and we became great friends.

As the campaign grew, so did my responsibilities. Eventually, I found the campaign an office and became volunteer coordinator—my first paid job in politics (with an impressive monthly salary of $900!). Among other things, I organized the drivers, baggage handlers, and go-fers who helped coordinate the candidate's visits to Los Angeles. These were heart-thumping occasions for me, as I got a chance to actually see the former vice president, the man who might be president, and to experience—for the first time—the rush of being around the motorcade, the Secret Service, the television cameras, the crowds, and the bustle of important and purposeful aides, forever in search of a telephone (in those pre–cell phone days) so they could call other important and purposeful aides to discuss important information with a strong sense of purpose. It was heady stuff.

In order to get as close as possible to the center of the action, I reserved for myself the all-important job of staffing the Staff Room, the hotel suite where members of the traveling entourage would gather. I made sure they had plenty of phones (and Diet Cokes) and that their messages were all efficiently placed in the envelopes bearing their names that I had carefully taped up on the wall. The traveling staff was a largely male outfit. From the moment Mondale's plane would land and staff would spill down the stairs and onto the tarmac, it was a virtual rolling scrum of white guys in dark suits. The women were scarce—but I was in awe of them. Maxine Isaacs was Mondale's press secretary, and Ann Stock—who would go on to become White House social secretary in the Clinton administration—was her assistant. And then there was Irene Tritschler. As Mondale's chief fundraiser in California, she wasn't technically part of the traveling staff, but when the candidate was in town, she always seemed to be standing at his side. I have crystal-clear memories of peeking out of the Staff Room to watch her glide down the hallway, surrounded by this phalanx of men, briefing Mondale as they walked to some Very Important Meeting a few doors away. It all seemed so big, so exciting—and suddenly, so within my reach. That could be me, I thought!

There were other women on that campaign who helped and encouraged and mentored me, most importantly Ali Webb, the state press secretary. Ali had taken a leave from her real job, as press secretary to then–Los Angeles mayor Tom Bradley, to work on the campaign. Though she was just a few years older than I, she had vastly more experience in politics, was smart as a whip and always great fun. We clicked immediately and started sneaking out of the office together for a chat, a quick lunch, or a drink after work. Later that summer, we drove up to the Democratic National Convention in San Francisco together. And during the general election, she hired me as one of her deputies in the press office. As I wrote press releases, talked to reporters on the phone (never for attribution!), and followed the campaign's every twist and turn in the news, I knew I'd found my calling. The press office was the place to be. And Ali helped me learn the skills I would need—and showed me, just by being who she was, how a woman could succeed as the voice, and often the face, of an important pubic figure.

Of course, the most visible woman on the Mondale campaign was Geraldine Ferraro, the first woman ever to run for office on a national ticket. I'll never forget the morning the news broke. I was temporarily living with my parents, and my dad called out, "Mondale picked Ferraro!" I confidently told him that was impossible, since Ferraro was no longer on the short list. As the volunteer coordinator of the California campaign, I was obviously in a position to know. "I just heard it on the radio," Dad said. "They announced it a little while ago." Disbelief quickly turned to jubilation. A woman, I thought. He picked a woman! To more seasoned observers, it was clear that Mondale's choice was risky. While Ferraro had obvious political gifts and she brought tremendous energy to a struggling campaign, her gender guaranteed that she'd get a more thorough going-over than the only copy of *Playboy* magazine in the boys' cabin at summer camp. To begin with, as part of the selection process, Mondale had invited candidates of every race and gender to his home in Minneapolis for interviews. Before retreating into the house, he made each of them run a gauntlet of television cameras in his driveway, turning the visits into a kind of Noah's Ark spectacle and fueling charges that he made his decision based on politics rather than qualifications. (Politics! In the vice presidential selection process! Surely you jest!) What's more, Ferraro was a former prosecutor and three-term congresswoman from Queens, New York. Her résumé was solid but unspectacular, making it inevitable that she'd be fighting charges she was unqualified from the get-go. And finally, since the country had never experienced a woman on a national ticket, it had never faced the prospect of a First, or in this case, Second Man. Ferraro's husband, John Zaccaro, became the subject of a thorough, and unflattering, series of investigations. By the end of the campaign, Geraldine Ferraro had become as much a liability as an asset.

Still, her mere presence on the ticket was a watershed moment for me. As my friend Ali and I drove to the convention, we were sure it would change not just the campaign but the world. During that miraculous week, I worked as a runner taking messages, packages, and information from the campaign offices in the bowels of the convention center to the various delegates and delegations on the convention floor. The work was hardly challenging, but it was fun. And most importantly, it meant that I had a floor pass. I realized that if I played my cards right, I could watch the big speeches from the front row. On Wednesday evening, just before Ferraro was set to accept her nomination, I helped distribute small American flags to the delegates, staffers, and hangers-on who were increasingly crowding onto the floor in anticipation of an historic moment. All the while, I plotted my own strategy for finding a great spot, right up front, to wave my flag and to revel in the grandeur of it all.

Moments before Ferraro appeared on stage, I squeezed myself into a space at the foot of the podium. Then suddenly, there she was: white suit, Dorothy Hamill haircut, energy and optimism bursting from her trim frame. (At forty-eight, she was just two years older than I am as I write this.) The crowd went crazy, those little American flags like buzzing bees' wings in red, white, and blue. I was so close that when Ferraro stepped up to the microphone, I could only see her head above the towering podium. And as she spoke, I was sure my life would never be the same. In some ways, it hasn't been. Being part of that moment, breathing in the sense of possibility, gave me faith that as a woman, I could play a role in national politics. After Ferraro's rough treatment at the hands of the press and her Republican opponents, I never assumed it would be easy. And in that, I haven't been disappointed. But I also assumed it could happen. And that made everything else possible.

When I talked to Geraldine Ferraro for this book, she played down the impact of her achievement. "These days, when I call someone's office," she told me, her wit and Queens accent as sharp as ever, "and the person answering the phone is under thirty-five, it's always the same. 'That's Ferraro. F-E-R-R-A-R-O,'" she laughs. And I laugh. But there are millions of us who are over thirty-five.

When More Is More

Virtually all of the women I spoke to for this book talked about the importance of role models. Kansas Governor Kathleen Sebelius grew up in a political family, but she never saw herself as a candidate when she was young. Although her father had been elected to the U.S. House of Representatives, and then governor of Ohio, there were no women out front in his campaigns. "The women licked the envelopes and answered the phones and did the backroom stuff, organized stuff," she said. "Very few women were ever involved in the policy discussions." But when the state representative from the Wichita district where she lived decided to run for another office, leaving her old seat vacant, something changed. "That was the first time I thought—I had lobbied the legislature, I did a bunch of stuff—and I thought, 'Well, hell, I can do this.' . . .

"Kansas has a very long, very rich history of women in leadership, and I think looks at women differently," Sebelius told me. Founded as a "free state," there was strong support for the abolitionist and women's rights movements early on. "Women could own property here long before they could own it in other parts of the country. Women voted here twenty years before they had the franchise nationally. There were all-women governments before women could vote. So there's a real credibility, long-standing credibility of women in government, women in leadership positions, that doesn't exist in some other states."

Clearly, role models do more than allow women to imagine themselves in a series of bigger and better roles. They allow other people—both men and women—to adjust, if that's the right word, to women in positions that have traditionally been filled only by men. And once they do, the road ahead becomes a little easier. Sebelius became not the first, but the second woman elected governor of Kansas; Joan Finney had been elected to the same job eight years earlier. The state also sent a woman, Nancy Landon Kassebaum, to the U.S. Senate. Once a state has elected a woman to the legislature or the statehouse or the Senate, it is more likely to elect another one. And another one. California and Maine both have two women serving in the U.S. Senate. Louisiana has a woman in the U.S. Senate and, until recently, had a woman governor. And Washington has two women in the U.S. Senate—and a woman governor. Sometimes more really is more.

Certainly that's been true in the film industry, where women have made astonishing gains in the past twenty years. Sherry Lansing told me that when she started out, women's options were very limited; they had lower-level jobs, like script reader or story editor, with little chance of advancement. "And I did say something like, 'There will never be a woman head of studio in my lifetime.'

"I had to eat those words," as she loves to tell it, when she tore through the celluloid ceiling, becoming the chief of Paramount in 1986. By 2005, five of the seven major studios had women at the helm. "And it's no longer a front-page story," she says. "Women are totally accepted in show business now."

Why? Lucy Fisher, a producer, believes that it's at least in part due to Hollywood's "immigrant, outsider ethos." The industry's first entrepreneurs and impresarios were immigrant Jews who often couldn't find work in more established fields. In other words, it's never been about pedigree or looking the part, at least not in the executive suites. "Here," she said, "if it makes money and you're a gorilla, you're in." In the absence of rigid expectations about how someone in a particular position is supposed to look or behave, people—both men and women—are judged by their talent, by whether or not they produce, a standard that has worked well for women. And the more women see other women succeeding, the more they are drawn to the business, and the cycle of success continues.

That doesn't mean change will come quickly in every area, or even in most; it hasn't and it won't. In too many fields, from politics, to business, to the faculties of most universities, women are still woefully underrepresented. "The notion that you can't see anybody in a university who looks like you is, I think, a terrible signal to be sending a student body that is 50 percent women," Shirley Tilghman, the president of Princeton, says of often disproportionately male faculties. But, she notes, women are finding greater success in college and university administrations. Many of the country's most prestigious universities, including Harvard, Princeton, Brown, and MIT currently have female presidents. The power of that example is palpable. "I can't tell you how many times I've had a really distinguished woman member of the faculty come into my office in the last five years and say, 'I want to be a dean, I want to be a president, will you help me get there?'" Tilghman says. "They see that women are succeeding in these kinds of jobs, and it's very attractive to them. These are not women who are failed academics, by the way. These are women who are at the top of their game. They're actually reaching a point that I think I reached, which is this: Okay, I'm at the top of my profession and I can stay there—or more likely start on the other side of the hill—or I can find another challenge. And it's great."

Into the Bathroom with the Boss

Role models play different, well, roles, depending on where we are in our lives and what we need. After Walter Mondale and Geraldine Ferraro got crushed (losing forty-nine out of fifty states), I was surprised (there's no believer like a young believer!) and heartbroken. But I wasn't discouraged. Before long, I was working for a California state senator as a field representative. I loved my boss, Art Torres (now chairman of the California

Democratic Party), but missed working with the press. So the following year, when my friend Ali Webb offered me a job as assistant press secretary to the mayor of Los Angeles, I jumped. I was twenty-three years old and on my way, thanks to the women in my life, and I would spend the next ten years working in a press office.

At every stage, there were women who helped and encouraged me. That's not to say there weren't plenty of men; there were. Mayor Bradley not only inspired me by the extraordinary example of his own life, but he treated me with kindness and respect; he always made me feel that I had something unique to contribute. Bill Carrick, who ran Dianne Feinstein's campaign for governor in 1990, was the best campaign manager I ever worked for. And of course, President Clinton gave me an opportunity that, though complicated (see chapter 1), changed my life. I will be forever grateful. There were other men, too; lots of them. And I don't mean to minimize their impact on account of their gender. That would be sexist, dammit. And it would be wrong. But the women played a different role. They helped me imagine the possibilities—and sometimes the perils—that lay ahead for me.

It was exciting when Susan Estrich took the reigns of the Dukakis campaign in 1988, becoming the first woman to run a national campaign. But her experience—she was given responsibility for running the day-to-day operation without the requisite authority—became a cautionary tale. Still, Susan survived and thrived, and thanks to her example, even in the most discouraging of days, I was pretty sure I could, too.

And then there was Dianne Feinstein. I went to work for Dianne in the spring of 1989, when she was gearing up to run for governor of California. She was already something of a political celebrity: She'd become mayor of San Francisco when her predecessor was assassinated, and had gone on to serve two successful terms. What's more, she had been one of the potential candidates for vice president paraded down Walter Mondale's driveway a few years earlier. She was tall, elegant, and authoritative; when she walked into a room, you knew she was there. Every politician I had worked for before—or since—was a man. I was excited about the possibility of working for a woman—and of helping to elect the state's first female governor. I realized how different things would be early on when Dianne ducked into the restroom at an event, and rather than cooling my heels down the hall as I always had in the past, I breezed right in behind her. For the first time, I was the one with special access and information, those magic ingredients of power.

Dianne didn't win that race. But she fought hard against Pete Wilson, who was a U.S. senator at the time. And in spite of the stakes, the campaign never turned into the kind of nasty, personal, and destructive mud bowl that we've become all too accustomed to in recent years. She survived with her reputation not just intact, but enhanced. Two years later, she was elected to the U.S. Senate, and she's been reelected three times since.

Working for Dianne, I watched another woman survive the rough and tumble of big-time politics. What I'd seen Geraldine Ferraro do from a distance, I watched Dianne Feinstein do up close. She didn't complain much about the double standard—except when it affected her family. Like Ferraro, Feinstein saw her husband's business deconstructed deal by deal, and his motives impugned. Nothing did more to bring out the fight in her. But most days, she campaigned with grace and confidence. I remember one episode in particular, early in the campaign. Following a series of disagreements—and a hysterectomy that sidelined Feinstein for several weeks—her chief consultant quit. He faxed an emotional and unflattering letter to every reporter in the state (before he told Feinstein), questioning whether Feinstein had the "fire in the belly" to win. (She would later joke that she thought she'd had the fire *removed* from her belly.) The initial news accounts said that the consultant had fired the candidate, and the campaign went into freefall, as most of the political operatives and observers predicted that Dianne would collapse in a heap and quit. Wrong! She took stock of the situation: She wanted to be governor, she was qualified to be governor, and she had as good a shot as any Democrat at winning. There was no stinking way she was going to let anyone else make that decision for her. So she rallied her troops, hired two fabulous new consultants, and within a few months, she was leading the field. She won the primary and came close to winning the general election.

So she ran, and I learned. I learned that success and setbacks are inevitable—and I watched Dianne handle both with magnanimity. I learned, again, that women face an additional set of obstacles—and I saw Dianne overcome many with determination and discipline. And I learned that women's life experiences—the professional and the personal—prepare them to lead. Not only had Dianne come to office in the aftermath of an assassination, but her family had survived a bomb planted in a window-box outside her daughter's bedroom that mercifully failed to detonate. She calmed the city in the aftermath of the Jonestown Massacre, recognized the beginning of the AIDS epidemic, and balanced the city budget. She raised a daughter and buried a husband. She was understandably impatient with questions about whether women were tough enough to lead. Her life was both trial (she passed) and testimony (she inspired). And once again, my horizon got bigger.

MAKING IT UP

So I was helped along by the women who came before me at virtually every stage of my life and career. But for others, there just weren't any women doing what they wanted to do when they wanted to do it. They had to make it up as they went along. And with a little help—sometimes from other women—and a lot of pluck, they created their own futures. For them, the focus was more often on being, rather than seeing, role models.

When Jane Goodall went into the jungle to study chimpanzees in the early 1960s, she was the first and, for a time, the only woman in her field. But Louis Leakey, one of the world's leading primatologists, believed that women were particularly well-suited to the painstaking work of observing animals in the wild. Although Goodall was young and relatively inexperienced, she had worked with Leakey, and he believed she was just the person for the job. What she discovered rocked the world—and inspired a generation of scientists. Two women in particular, Dian Fossey and Biruté Galdikas, contacted Leakey begging for similar jobs; he sent Fosse to observe gorillas in Rwanda (where she was eventually murdered and became the subject of the film *Gorillas in the Mist*). And he sent Galdikas to study orangutans in Borneo. Today nearly two-thirds of the world's primatologists are women, a reality unimaginable without Goodall.

"After every single lecture, people come up to me and say, 'I did what I did because of you.' And they're mostly women," Goodall told me during a conversation on a recent visit to Washington. "And girls around the world have said to me again and again—here, in China, and in Europe—they've said, 'I didn't think a girl could do that kind of thing. But then I read about you and realized that if you could do it, I could do it, too.'"

In the absence of actual living, breathing examples, countless women have turned to books. Certainly, fiction has been a rich source of inspiration. As my daughter, Kate, and I have worked our way through all seven *Harry Potter* novels, I've been delighted to see how much she identifies with Hermione Granger, the occasional know-it-all who is also so smart, diligent, loyal—and absolutely indispensable.

And then there's biography. "It was in college, through my books, that I met Frances Perkins, the first female cabinet member," says Alexis Herman, who would follow in Perkins's footsteps, herself becoming labor secretary. Likewise, Vera Rubin, an astronomer who discovered dark matter, didn't see many women in science while she was growing up. But she came across some in the library. "I found a biography of Maria Mitchell, a female astronomer who discovered a comet in 1847, and that's about when I made the decision to become an astronomer." Now all four of her children are scientists.

Similarly, Senator Kay Bailey Hutchison has been first more times than you can shake a mesquite stick at: the first woman television reporter in Houston, the first Republican woman elected to the Texas state legislature, the first woman to be elected to represent Texas in the U.S. Senate. She didn't have role models in the traditional sense; no women had held the jobs she wanted. But as a girl, she read biographies, lots of them. "I'll never forget when I was in the sixth grade, I had already read every biography in our school library and couldn't fulfill my biography requirement because I had already read everything. Now, it was a small library [LaMarque, Texas, population 15,000], but I think that biographies did encourage me to think that everybody went through hardships, but if they had a real determination, they did great things." Hutchison said virtually all the biographies she could find were about men—Betsy Ross was the only woman—"but I just didn't think of myself as different from them, from the men," she has said. (It's worth noting that Hutchison has now written three books, and all tell the stories of successful women.)

Things started to change when Hutchison went to law school at the University of Texas. She was one of only seven women among the 236 graduates in the class of 1967, and like so many women of her era, she got the what-are-you-doing-here-and-why-are-you-wasting-a-spot-that-might-have-gone-to-a-man treatment. Although she persisted and graduated near the top of her class, Hutchison couldn't get a job as a lawyer; no firm in Texas would hire her. But with a little help from a couple of women, she found her own path. "On a lark," she interviewed for an on-air position at a local television station. Oveta Culp Hobby—the former U.S. secretary of health, education and welfare; a former commanding officer of the Women's Army Corps; the then-chairman of the board of the Houston Post Company—owned the station, and she decided to give her a shot. According to Hutchison, Hobby later told the station manager, "Well, it would be hard for me *not* to want to hire a qualified woman and to be the first in Houston to do it."

After Hutchison had covered the state legislature for a few years, the chair of the Harris County Republican Party in Houston—also a woman—asked her to run for a newly created seat in the state legislature. At the time there were just four Democratic women in the house, and no Republicans. But Hutchison was intrigued by the possibility. She ran, and she won. "So a woman gave me that first chance [to work in television] and a woman asked me to run for the legislature," she says.

While she's quick to remind me that as a U.S. senator she represents all the people of Texas—not just women—she's conscious of being a role model. And she believes that when women succeed, it not only encourages other women to pursue challenging careers, but convinces the world that women are ready. When I ask her whether she thinks the country is ready to elect a woman as president, she doesn't miss a beat. "I do. I do. I do think so." Later, she adds, "You look at what people are going to look at for president and there is nothing different—other than that we had to fight harder. But they know that we've had the experience. We've been in foreign policy, we've led a state, we've been a CEO . . . You know, maybe I'm prejudiced because I don't think there's a difference. But I think that people will vote for a woman for president or vice president if they like what she wants to do for the country and they know she's had the experience and has shown that she can do the job."

I hope Senator Hutchison is right and that the country really is ready to elect a woman president. I'm not sure. But as Hillary Clinton's campaign steams along, we're getting ready to test the proposition for the first time. So far, the public seems to believe that she's qualified, a hurdle that most women who've toyed with the idea of running in the past haven't been able to clear. After one in an endless series of early debates among the Democratic candidates, Doug Burns, a columnist for the *Iowa Independent*, wrote: "No longer should the question be can a woman be president? With her fourth in a series of crushing Democratic debate performances, the question everyone in America should be asking is: What have we been missing by eliminating more than half our population from the application process for this job since the late eighteenth century?" What indeed. Still, I'm convinced that Hillary has been tested in ways that her male opponents haven't. Firsts always are. The state of her marriage, for instance, will be an issue. Yes, male candidates get asked about their marriages, too, especially if they've had more than one. But the answers don't matter very much. For Hillary, they are critical.

To be sure, gender isn't the only issue that Hillary faces. After all, Hutchison said that while Hillary is qualified, there's no way she would ever vote for her; ideology trumps biology for most voters. But gender will always be an important part of her story. And the first woman elected president, whoever she is, will become the most powerful *person* in the world, an inspiration to generations of girls, a monument to the idea that change is possible. She'll become the mother of all role models.

For decades, women have argued that role models and mentors would help create a mighty river of opportunity for women. And while women are increasingly visible, progress has been agonizingly slow. There are nowhere near as many women in the top ranks of most professions as people hoped or expected a generation ago. But in other ways, progress has been swift, if not always steady. In 1967, married women in Texas couldn't control their own property or their wages. *1967!* Now the senior senator from Texas is a woman. So are the president of Chile, the chancellor of Germany, the U.S. secretary of state, the CEO of eBay, and the president of Harvard. And potentially, the president of the United States. Together these women—and the thousands of women closer to home who are not only the nurses, teachers, and nuns, but also the doctors, lawyers, principals, entrepreneurs, and engineers—are certain to inspire young girls, and even a lot of boys, broadening their horizons, expanding their ideas about what's possible, whatever their politics.

As she traveled around the country in the early months of her campaign, Hillary Clinton said she was struck by two types of people who so often appeared at her events: women in their nineties who want to help make history and parents of young daughters.

"As I go by shaking hands and meeting people," Hillary said in a speech in Ames, Iowa, "I often hear a dad or a mom lean over a little girl, and say, 'See, honey, you can be anything you want to be.'"

Seeing really is believing.

Chapter 10

Reaching Critical Mass

"The thing women have got to learn is that nobody gives you power, you have to take it."

—Roseanne Barr,
comedian, actor, and writer

I opened my last briefing as White House press secretary with two David Letterman–style Top Ten Lists. The first was of the president's accomplishments from the year that was winding down, and the assembled scribes groaned as I recited such memorable milestones as the Interstate Banking Bill. The second was a list of things I wouldn't miss. Helen Thomas was both number ten and number two. She delighted at being a thorn in the White House's side, and feigned irritation that she hadn't been number one. The momentarily lighthearted mood turned a bit more serious, as reporters started peppering me with questions about a $4 million advance that then-speaker Newt Gingrich had received for a series of books he was planning to write.

"I haven't discussed this with the president, and I don't know what his thoughts are on this book," I said.

"Can you take the question," asked a reporter, meaning would I go ask the president and then pass his response on to the press.

"Sure," I started to answer, when the room burst into applause. Confused, I looked to my right, and was utterly stunned to see the president striding toward the podium.

"I thought I should come in and get you out of hot water, since that's what you've been doing for me for years," he joked.

He jousted a bit with the assembled reporters, then turned valedictory, offering me his parting gift. "I just wanted to come in here and say in front of all of you how very grateful I am for everything Dee Dee has done for me since long before I became president, starting in our campaign. I reminded her that the first trip we took together was on a little bitty airplane, and I fell asleep, which was some sort of comment about how helpful I would be in answering difficult questions.

"And we've had a wonderful professional relationship; we've had a good personal friendship. I think she is one of the best people I have ever had the privilege of working with. And I'm really going to miss her."

But before long, the beasts (as I lovingly called the reporters in the briefing room) had had enough of what the president thought about me; they wanted to know that he thought of Gingrich's book deal. He chuckled, amused by the predictability of it all. And a minute later, he was out the door.

I was delighted by the president's cameo—and grateful for his words. (George Stephanopoulos and Mark Gearan had organized it, and I still love the picture of them peering around the corner as the president entered the briefing, grinning broadly at my surprise.) The previous months had been brutal for me, as I had struggled to both re-create my job—and hold on to it. In the end, I settled for the former—and for the satisfaction of knowing that I had done my best to protect the president, to act honorably, and to draw a line in the sand about how women in similar positions should and would be treated.

The beginning of the end came in July. After seventeen turbulent months as president, Bill Clinton announced plans to shake up the West Wing, starting with the appointment of a new chief of staff, Leon Panetta.

The night of the announcement, Panetta appeared on *Larry King Live* with his predecessor, Mack McLarty. When King asked them specifically if I might be leaving, they bobbled the question, leaving the impression that my future was in doubt. Although Panetta called later that evening to apologize and to say that he had "full confidence" in me, his comments—whether intentional or not—had made my fate an open question. I knew I was in for a bumpy ride.

Across the next couple of weeks, the questions continued to swirl, as I tried to convince the president and Leon that they needed a stronger press secretary, someone with more access and more authority than I'd been given. I urged both of them to fix the job, with me in it. But also made clear that if they wanted someone else, I'd be gone in a heartbeat.

At one point, Leon called me into his office, a flowchart mapping out the new structure for the press operation spread across his conference table. At the top was an uber–press secretary, a new position that had yet to be filled. It would go to a seasoned pro, someone with deep Washington experience who could pull the press and communications operations together. Below that was a box called "Traveling Press Secretary"; that person—me—would handle the daily briefings and travel with the president.

"It won't work," I told him, a bit surprised by my boldness. "And I won't take that job. You've got to have one person, and it's got to be clear. You can't split the job up." We agreed to keep talking.

Toward the end of September, Mike McCurry, who was then chief spokesman at the State Department, met with Panetta; when word of the meeting leaked, the Reuters News Agency ran a story under the headline "State Department Aide to Become Clinton Spokesman." I was livid, and I marched down the hall to have it out with Leon. I'd been a loyal member of the team; surely I deserved better than to read about my replacement on the wires, I told him. Panetta was unsympathetic; he hadn't offered anyone my job, he said—but he reserved the right to talk to whomever he pleased. Finally, Leon told me I needed to meet with the president.

A few hours later, I walked into the Oval Office. The president was sitting behind his desk, and I slipped into one of the hard-backed chairs facing him. I told him how disappointed I had been by the way things had been handled, that I felt like I'd been left to twist in the wind for months. Clinton said he knew it had been tough for me, that he'd never meant for that to happen and that he was genuinely sorry. Then we talked about what to do. "Fix the job, and give me a chance to do it," I said. "And I'll leave by the end of the year." That would reestablish the authority of the press secretary in a way that would better serve him, I explained—and give me a chance to prove that I could do it, however briefly.

I knew the president didn't want me to leave—not like this. But I think he knew I couldn't stay, not indefinitely, not now that things with Leon had gone so far down the track. And all the reasons my status had become an issue in the first place were still there: The job as I'd been given it was compromised—and that had undermined me and fed doubts about my effectiveness. He wanted to give me a chance to get it right, but in many ways, it was too late. So he agreed to my terms—though neither of us was happy about how it was going to end.

As I walked from the northwest gate to the West Wing entrance the next morning, a gaggle of cameras followed me. Overnight, the story of my meeting with the president had leaked, and news of my "promotion" had become the early lead to the staff-change story. While I was gratified that the president had given me what I'd asked for, I also knew that my reprieve was temporary. By Christmas, I would be gone.

The last months were both satisfying and difficult. I got the increased access, the traditional office, the higher rank and salary. After nearly two years, I finally felt I had the tools I needed to do my job. But the price I paid to get them was agreeing to leave.

In addition, Panetta became one cold customer. In mid-December, I was summoned to his office. "Be ready to leave by the end of the month," he said, as if he'd been looking forward to the moment for three months. He hadn't needed to tell me.

As it turned out, December 22 was my last day. After my final briefing, I went back to my office, for the West Wing version of sitting shiva. There were stories and signed photos, cake and champagne. Friends and colleagues—some from the earliest days of the campaign, others from our time together in what would turn out to be the first quarter of Clinton's presidency—came to say good-bye.

Later that evening, as I left the White House—for the last time as the president's press secretary—I knew my life would never be the same. In some ways, the job hadn't been everything I had hoped. In others, it was more than I could have imagined. And for better and for worse, I would always be the first woman.

BETTER ENDINGS

I may have been the first, but I'm no longer the only. In September of 2007, President Bush named Dana Perino as his press secretary, making her the second woman to join the little club of presidential spokesmen. I was thrilled for her. More importantly, I was confident that she wouldn't confront many of the same issues that I had. She'd already been at the White House for several years, had been her predecessor's chief deputy and knew her way around Washington. She took over late in the president's tenure, rather than early. And nearly fifteen years later, it would be a whole lot more difficult to give a woman more responsibility than authority in that particular job. Different opportunities had created different expectations, and Dana's story will have a different ending than mine did.

I actually got the chance to rewrite parts of my story—for television. And it was a kick. A few years after I left the Clinton administration, I got a call from Aaron Sorkin, the gifted writer who had already had several hits on Broadway and in film, including *A Few Good Men* and *The American President*. He was now turning his talent to television, developing a series that would become *The West Wing*. Would I be interested in joining the show as a consultant, he asked? I jumped at the chance. For the next six years, I worked with the writers, producers, directors, and actors to help add a bit of verisimilitude to the show's characters and events. It was a rare opportunity to learn about another business—

television drama—from the inside, and to meet and work with a wonderful and talented team. I loved it—and often joked that if I'd known there was a job where you were *supposed* to make things up, I never would have wasted all those years in politics.

Among other things, I helped shape story lines based on what I had seen and experienced during my time in the "real" West Wing. And it was in working through certain memories that I got to take things that happened to me—and make them come out better for my TV alter ego, the press secretary C. J. Cregg.

There were two episodes in particular that were especially satisfying. In one, C.J. is left out of the information loop, as I was, as the president plans to take military action against a foreign country. But unlike me, she finds out about the strike before it happens, and helps the president and the administration successfully communicate its actions to the public. In the other, C.J. has to have a root canal (a plot device that Aaron created simply because Allison Janney, the wonderful actress who played C.J., is such a brilliant physical comedian). After the procedure, C.J. can't brief the press, so her colleague Josh Lyman takes the podium in her place. His clumsy and imprecise remarks create a monetary crisis—driving home the self-serving point that briefing is harder than it looks.

Not long ago, after I made a speech on a college campus, a young woman came up and said to me, "Oh, so you're like C. J. Cregg."

"No," I said, mindful of C.J.'s slightly altered circumstances. "She's like me."

Television isn't the only place where women are rewriting history. Increasingly, they're taking the experience of their predecessors—the men and women who came before them—and creating their own story lines, with new opportunities and better endings.

Several years ago, Frances Kaiser, the sheriff of Kerr County, Texas, arrived at the home of a man she had known for many years. Going through a divorce and distraught, the man had a gun and was threatening to kill himself. "I knew that he really didn't want to do that. I knew him," she said later. "I went in and I told him that he needed a hug. And I put my arm around his shoulder. He was sitting in a chair with the gun down between his knees, and when I touched him and put my arm around him, he just wilted, he just relaxed. And the other officer that was with me took the gun.

"I would probably never do that again, you know, after I thought about it," she explained. "But I did it from the heart. I did it with compassion. I did it because I knew that he really didn't want to do that. But he was crying out for help."

Kaiser calls it tough love. "We have to make people accountable for their actions, but we also have to be there to support them. And I think we as females have that trait, where maybe males do not have that trait. And I take pride in saying that because I feel that way about myself, and I feel that that's probably why I am where I am today."

It's easy to imagine a different ending to that particular story. But one woman—acting on instinct and with compassion—brought a whole range of new options to an all-too-familiar situation. And she created a better outcome for everyone involved.

The Thatcher Question

When I told people I was writing a book exploring the ways women have changed public life, the most common response—particularly from men—was: Well, what about Margaret Thatcher? Isn't she proof that women in power don't change anything? After all, the Iron Lady, as she was dubbed by the Soviet press after treating that country to one of her rhetorical beatings, governed like a man. True, she was a no-nonsense conservative who believed that responsible individuals were the building blocks of a responsible society, that government should do less so that people could do more, that individual initiative alone could restore the glory of Britain. Rectitude, not compassion, was her hallmark. Among the more memorable elements of her decade-plus tenure as prime minister were the war in the Falkland Islands, her standoff with labor unions, and her friendship with Ronald Reagan. None showcased the softer side of politics that people like me often expect—or at least hope—to find when there are more women.

Neither Thatcher, nor the feminist leaders of the day, considered her one of their own. The writer Jill Tweedy once said, "Mrs. Thatcher is a woman but not a sister." And Mrs. Thatcher made it clear she didn't like "strident females." "I like people with ability who don't run the feminist wicket too hard. If you get somewhere, it is because of your ability as a person, not because of your sex."

The movement, as such, was clearly not her thing. And yet, she saw clear differences between men and women. "In politics, if you want anything said, ask a man. If you want anything done, ask a woman," she once famously said. And if she believed that her gender didn't account for her success, she also seemed to believe that the particular qualities she brought to the job *because* she was a woman were damned important. "I've got a woman's ability to stick to a job and get on with it when everyone else walks off and leaves it," she once said. Clearly, she believed that women had much to contribute—and that the increased presence of women would make things better. "The woman's mission is not to enhance the masculine spirit, but to express the feminine. Hers is not to preserve a man-made world, but to create a human world by the infusion of the feminine element into all of its activities." In other words, women, by their very nature, could change the world. And clearly she didn't mean by simply tending hearth and home.

Some might argue that Thatcher's presence did little to infuse the global political sphere she occupied with the "feminine element." But is it really fair to expect one woman to change a system created by men and for men over countless generations? She faced all of the predictable obstacles that women had long faced. When Thatcher led her conservative party to an impressive victory in May of 1979, becoming the first woman to head a modern Western government, the late R. W. Apple of the *New York Times* wrote about some of her perceived weaknesses. "Her voice and her manner reminded many voters of unfondly remembered schoolmates, and many of those who liked her policies could not bring themselves to help her become Prime Minister."

When Thatcher became prime minister, the vast majority of members of Parliament, of her government, of the leadership of the Conservative Party, and of the power structure of the UK and the world were men. With the possible exception of the queen, who got where she was by the happenstance of birth ("the lucky sperm club," as a friend of mine would say), she was without female peers. Yet somehow, despite being vastly outnumbered in a world where she had to adopt a long list of male behaviors to succeed, she was supposed to change it all single-handedly.

That's not to say that Margaret Thatcher couldn't have done more. She appointed just one female minister during more than a decade as prime minister, a fact that frustrated women in both the Conservative and Labour movements. As Patricia Hewitt, a minister in Tony Blair's cabinet, told the BBC, "Margaret Thatcher damaged women's place in the workplace, undermined families and communities, and did nothing for women in public life. It was a wasted opportunity on a gargantuan scale."

But is it really fair to say she "did nothing for women in public life?" When she became prime minister in 1979, there were only twenty-seven female members of Parliament, and when she left Downing Street in 1990, there were forty-three. Clearly, there were other forces at play, as women assumed more political power in many parts of the world. But Thatcher obliterated any questions about whether a woman could be tough enough to lead.

Kim Campbell, the first woman to serve as prime minister of Canada, says Thatcher "created a constituency for women leaders where one hadn't existed before." As she traveled across the country during her tenure, Campbell said, "people were very excited about having a woman as prime minister. Little old men would come up to me and say, 'Oh, you're going to be our Maggie Thatcher.' Well, I wasn't really much like Margaret Thatcher, but she had created this image of a strong woman leader."

More Than One

Not long ago, Xerox began planning for succession. On the brink of bankruptcy just a few years earlier, Xerox had fought its way back to financial health, and the powers-that-be wanted to avoid some of the transition mistakes that had helped put them in such a precarious position in the first place. But almost everything else about the process was unusual.

First, the board of directors had such confidence in the CEO who had engineered the turnaround and in the longtime heir apparent that it gave the two virtually unprecedented authority to work out the details themselves. The players—both considered straight talkers—discussed their challenges with a rare level of openness in the hopes that corporate America might learn something from their process. And when Ursula Burns takes over from Ann Mulcahy, it will mark the first time in Fortune 500 history that one woman CEO will succeed another.

In a series of conversations with *Fortune* magazine, Mulcahy and Burns didn't talk much about the fact that they were women; they didn't have to.

It showed in the metaphors they used. When describing the challenge of relinquishing power, Mulcahy said it's difficult to give up both "clarity of control" and "the incredible pull of being needed all the time. It's like your kids growing up, I guess, right? 'Oh, I'm not the center of the universe anymore.'"

It was clear in their approach to leadership. "At this level, it's not about 'I told you to do it, now go do it,'" Mulcahy said. "Some people do their own thing. Some people have to be hugged and loved. To her credit, Ursula gets it."

And it was obvious in the way they communicated with each other. "We talk about everything," Burns said. "Movie stars, the famous people, business. You know, our kids, big time about our kids."

A generation ago, it would have been hard to imagine they could talk like that, think like that, lead like that—and still make it into the executive suite. Not once, which might have been called a fluke. But twice. Back to back. Mulcahy and Burns didn't adopt all the tribal customs of men. Instead, they wrote their own story; they created their own ending.

If one or two women can change an outcome, how many women does it take to change the culture, to create an environment where women no longer have to conform to male norms? When do we reach critical mass? (The term "critical mass" is borrowed from physics, and it refers to the smallest amount of fissile material required to start an irreversible chain reaction.)

Sometimes "a few" just isn't enough. During my years in the White House, there were a respectable number of women in top positions. But despite our increased visibility—and the certainty of some shared experiences—the women of Clinton's White House didn't really band together to share information, strengthen alliances, or just seek relief from the frustrations of our jobs, not all of which were gender-specific. Christine Varney, who served as the administration's first director of cabinet affairs, and I discussed this over lunch recently. We agreed that both of us could have benefited from spending more time with the other women in comparable, and at times isolating, positions. So why didn't we?

Some of the answers—there weren't enough hours in the day, we didn't always have overlapping responsibilities—are both true, and incomplete. Also at play was the sense that as women, we were already a little marginalized. Spending too much time with other women—or taking up the cause of women more broadly—would just push us that much further from the all-important center of power. Most women seemed to feel that in order to survive, let alone get anything done, they had to make common cause with the players—and far more often than not, the players were men. In hindsight, it seems clear that we were wrong, that having an occasional dinner together would have empowered us, or at least given us the chance to compare notes and confirm that we weren't the only ones. But at the time, something kept us from doing it; something led us to believe, perhaps unconsciously, that there would be a price to pay.

Of course, Hillary Clinton was an important player in that

White House. Not only was she the president's wife—and the first First Lady in history to have come to the role with a profession, a career, and a power base all her own—she was also a policy player and a political force, particularly during the years that I was there. Then, as now, her staff was largely a loyal cadre of women who were less likely to leak or to leave than those of us who worked for her husband. And then, as now, she had a large circle of friends, mostly women, outside the White House, whom she'd been collecting since her days at Wellesley. But her relationship with the women who worked in the West Wing was at times more complicated.

I, for one, never felt she was particularly supportive of me—or even sympathetic to the ways in which being a young woman made my job harder. In some ways, I understood that her challenges were bigger than dealing with me: In the fall of 1994, as I was trying to save my job, she was dealing with the collapse of health care (for which she was amply blamed), the Republican takeover of Congress, and a husband in political crisis. If the choice was changing the press secretary or changing the president, it was a no-brainer. And yet. It wouldn't have killed her to show some empathy, even if she stopped well short of coming to my aid.

That said, I think Hillary often felt that the president's staff wasn't particularly supportive of *her*. She believed that privately—and sometimes publicly—we were all-too-eager to throw her under the bus if it might improve her husband's political fortunes, however briefly. And sometimes she was right.

Still, Hillary fought hard for the priorities that had been important to her throughout her years in public life. She used her considerable power to advocate not just for health care reform but also for early childhood education, protections against domestic violence, and more women (and minorities) at all levels of government, among other things. And because those issues were important to her, they became important, *period.* Her presence made a difference. But her power was still derivative; she wasn't the boss. And the culture of the White House was vastly more the president's than hers; it was still a place where women were outnumbered—and too often out of the loop.

What would the White House be like if Hillary were president, if her power were rooted not in her marriage but in the Constitution? Time may tell. Her years as First Lady surely hold some clues. But more important, it seems to me, are her years in the Senate. I didn't work for her (or anyone else) there, but the record seems to suggest that when the power is her own, she wears it more lightly—and uses it to great effect.

The Right Number

"I happen to believe there is strength in numbers," says Alexis Herman. "I think the more of us you have in these hallowed places, then I think some of those barriers are going to naturally fall. I think the more we still have one or two women in key environments, it's still a disproportionate burden."

A recent study of corporate boards found it takes three women to really change the dynamic in the board room. A lone woman is often made to feel she represents the "woman's point of view," and can be left out of decision-making discussions and even social gatherings. Adding a second woman helps. But the "magic seems to occur when three or more women serve on a board together," the study concludes. Suddenly, women are no longer seen as outsiders, and their influence on the content and process of discussions increases substantially.

Women bring a "collaborative dynamic" that increases the amount of listening, social support, and win-win problem solving. They take on difficult issues, ask tough questions, and demand direct and detailed answers. And they bring different ideas and perspectives to the table, broadening the content of board discussions. In short, adding women makes the process more productive.

Of course, the magic number isn't always three. Would that it were. Different circumstances require different solutions. It may take more women to change the culture at a large aerospace company than a small accounting firm. A handful of women may be more powerful in an American state legislature than in the parliament of a deeply patriarchal African nation. There isn't a simple answer to the question, "How many?" But more often than not, it's still "more."

"In order for [men] to embrace you, in order for them to support you, you almost have to think like them," says Wangari Maathai. "And sometimes by doing that you are joining the bandwagon that you are trying to change. So I think that we need to continue working very hard to have more and more women get into leadership, and sometimes I think, well, maybe 50 percent is too ambitious. But if we had even a third present—a third—we might begin to make men think differently."

In fact, much of the literature suggests that when women make up roughly a third of legislatures or other elected bodies, they begin to have a significantly greater effect. But again, it's not always that simple. Studies show that women begin to have an impact much sooner, when they constitute something closer to a sixth. (Interestingly, the U.S. Congress is just now reaching that point.) But their power is usually limited. Even when they get closer to that elusive one-third, many of the men—who have often been in the legislature longer and hold more positions of power—are disproportionately influential.

Worldwide, only fifteen countries have national legislatures that are one-third or more women. But that experience shows that it does make a difference. In Denmark, Finland, and Sweden, more public resources are spent on programs like child care and parental leave that allow women to balance work and family. In South Africa, women have led efforts to make sure that government resources don't disproportionately benefit men. And in Rwanda, women have been able to change laws that once prohibited them from owning or inheriting property.

The changes ripple, like waves across a pond. Ritt Bjerregaard, the powerful mayor of Copenhagen (and former European Union commissioner), is trying to increase the number of women leaders in municipal government. And one of her strategies? Encouraging new fathers to take more time off when their children are born. The city offers up to forty-eight weeks of leave for new mothers and fathers, fourteen of them with full pay. But Bjerregaard found that while new moms were taking an average of 120 days, new dads were taking only fourteen. In an effort to nudge dads to take more time—and thereby spread the time away from work more equally between both parents—the mayor sends new fathers a package that includes a nursing bottle, bib, and diapers. And she expects them to use it.

Every country that has achieved critical mass has some form of proportional representation. In one widely used form, each political party submits a list of candidates to voters, and is then awarded representatives in proportion to the number of votes it gets. Parties and candidates who receive a block of votes—but not necessarily a majority—can earn seats, thereby increasing the representation of racial and ideological minorities and women. In addition, most of those countries also use some form of quotas and campaign subsidies.

THE MORE WOMEN SUCCEED

Of course, culture is also key. All of the Nordic countries—Sweden, Norway, Finland, Denmark, and Iceland—have long traditions of women in positions of authority. In 1906, Finland became the first country in Europe to give women the right to vote. (New Zealand had led the world in 1893.) In 1924, Nina Bang became Denmark's education minister. And in 1990, Elisabeth Rehn became Finland's defense minister, the first woman in the West to do so. But she immediately bumped up against an unusual obstacle: For generations, male political and business leaders regularly met in the sauna. "I certainly prefer to discuss matters with my clothes on," she made clear at the time.

All five countries are strong, mature democracies. All have national legislatures that are at least one-third women. And most have elected women as prime ministers, presidents, or both. Researchers say it's an extension of the Nordic region's agrarian and egalitarian history. "Rural communities were by and large matriarchal, and women worked alongside men," says Raimo Vayrynen, a political science professor at the University of Notre Dame.

It's worth noting that each year, the United Nations evaluates countries based on criteria such as life expectancy, income, and education. And for six years running, Norway has been named the best place in the world to live. Maybe it's not entirely due to the influence of women. Then again, five of the top ten countries have had women leaders in recent years. (The United States was ranked eighth.) What's more, the presence of women doesn't seem to have done much harm. The comedian Elayne Boosler pointed out recently that Norway has a vibrant economy based on oil and gas, mining, shipbuilding, fishing, and paper products. "Mary Kay cosmetics overtook none of these industries," she said.

More women have led governments in South Asia than in any other region of the world—but there, the culture is dynastic rather than egalitarian. For the most part, Asia's women leaders have been the daughters, wives, or widows of powerful men. Indira Gandhi became India's first female prime minister several years after her father died of a sudden illness. And many of the other women leaders, particularly the firsts, came to power in the wake of the hanging, shooting, or bombing of a male relative. The circumstances of these women's ascension generally gave them political legitimacy: Each was carrying the torch of the martyred hero. But most had little or no political experience, and the results have been mixed. Sirimavo Bandaranaike of Sri Lanka became the world's first female prime minister after her husband was assassinated. Corazon Aquino became the de facto opposition leader in the Philippines—and was then elected president—in the wake of her husband's assassination. And Benazir Bhutto's ascent to power in Pakistan began when her father was overthrown in a coup and then executed. She later became the first female prime minister of a Muslim country and served two nonconsecutive terms, but she left office both times amid allegations of corruption. After nearly a decade in exile, she returned to Pakistan in the fall of 2007, intending to run again for parliament. A populist and secular leader, she remained deeply controversial, and twelve days before the January 2008 elections, she was assassinated.

Even in countries without a history of dynastic tragedy, women more often have been chosen to lead by their peers in parliament than by voters in direct elections. In addition to Thatcher in the United Kingdom, past and present leaders such as Angela Merkel of Germany, Golda Meir of Israel, and Tansu Çiller of Turkey all came to power through their party structures. That's not to say that they didn't work hard to earn respect and authority; they did. Or that parliamentary elections aren't often a referendum on a particular leader; they are. Still, it's more difficult for women to win national elections, where they have to earn the confidence of millions of voters, than interparty contests, where hundreds—or even dozens—of their colleagues decide.

But even that's changing, and countries around the world are increasingly choosing women in national elections: Finland, Nicaragua, and Latvia have all elected women presidents. And Ireland, the Philippines, and Switzerland have each elected two.

In 2005, Liberia made Ellen Johnson-Sirleaf the first woman elected president of an African country, and in 2006, Chile made Michelle Bachelet the first woman president in South America. The following year, Argentina also elected a woman president, Cristina Fernandez de Kirchner. Significantly, Fernandez de Kirchner crushed thirteen opponents, winning a decisive victory. What's more, the candidate who finished second was also a woman, and between them, the two top contenders won more than 70 percent of the vote. With capable women poised to compete for high office in countries like Paraguay and Brazil, it's possible that more than half the residents of South America—a continent once synonymous with machismo—will soon be led by women, a reality unimaginable even a few short years ago.

The bottom line is: The more women succeed, the more women succeed.

And then there's Hillary Clinton. She became the twenty-first American woman to run for president of the United States—beginning with Victoria Woodhull in 1872 (nearly fifty years before women had the right to vote)—but the first to have a real chance to win. When the United States finally elects a woman president, it will signal a global sea change: For the first time, a woman will lead at least one country on every continent except Antarctica. And for the first time since humans split from our chimpanzee ancestors and marched out of the jungle toward a more civilized future, a woman would be the most powerful person on the planet.

Nothing will ever be quite the same.

Changing the World

I wasn't the first woman to be the first woman; nor will I be the last. We all stand on the shoulders of those who came before us, the countless others who stuck their necks out—and sometimes got their heads knocked off—for going where no woman had gone before. Not all of them were trying to advance the interests of the sisterhood. Still, because of them, those of us who followed have had more, different, and better opportunities. I know I have. And I owe a great debt to the women who blazed new trails in politics and government and journalism and business. And while progress hasn't always been as fast as many of us would have liked, the climb has been steady. And it continues. As *Fortune* magazine said in its annual feature listing the fifty most powerful women in America, "On the whole, we are optimistic; the trend is genuine and sustained progress that is good for women—and good for the companies they serve." Good, too, for the communities and the countries they serve. Good, in fact, for the entire planet.

"What I think is interesting is when women become present in greater numbers, they have greater confidence in being able to be women. And not only do they then speak in different voices—and perhaps we begin to see if there are any differences in their outlook—but they also change the culture," said former Canadian prime minister Kim Campbell.

Sometimes, it takes one woman; sometimes, it takes many. Almost always, I've found, when there are enough women in the room so that everyone stops counting, women become free to act like women.

It's then that we can eliminate double standards and accept that men and women are different—and that they bring a different range of experiences, skills, and strengths to public life. It's then that we can start to value women as much as men and to retool our institutions to fit the broad range of choices that women—and men—make. It's then that we can expand our definition of leadership—and of the language we use to describe it. It's then that we'll have more representative government, better schools, and more effective diplomacy. We'll have stronger communities and a fairer society. We'll be able to reduce conflict and build a better future.

It's then that we can take advantage of all that each of us has to offer. And it is then that women will rule the world. And when women rule, we will have changed the very definition of power. We will have changed the world.

ACKNOWLEDGMENTS

One of the great ironies of writing a book—in so many ways, a profoundly solitary endeavor—is that it takes a village (to paraphrase a famous woman, paraphrasing an African proverb). At least it did for me. And there are so many people to whom I'm grateful for pitching their tents around my campfire.

First and foremost, my love and thanks to my husband, Todd Purdum. He had the unenviable tasks of talking me off the ledge at the end of frustrating days, reading single paragraphs when I needed feedback *right now*, and pushing me forward . . . but not so hard that I felt like he was pushing me forward. And on way more days than not, I truly believed he didn't mind.

My children shared their mother with this project. And while they looked forward to the day when it would be done, they followed my progress with a mixture of pride, curiosity, and utter disbelief that anything could take so long. Connie Spaid and Cheryl Appleton cooked the meals, ran the baths, and read the bedtime stories when I couldn't, and I can never thank them enough for loving my children.

My parents, Steve and Judy Myers, eagerly followed my progress—and understood why we sometimes had to leave family gatherings a little early. My sisters, Betsy Myers and Jo Jo Proud, were cheerleaders (both, literally, in earlier lives, and figuratively!), rooting for me along the way.

Ali Webb continued to be the kind of friend who reads your draft not once but three times before offering just the right mix of sharp-eyed edits and unbridled enthusiasm. Marilyn Smith shared her wise insights, always keeping both eyes on my best interest. And Lee Satterfield listened to my stories, asked smart questions, and helped me "think through" the big stuff and the little stuff, day after day after day.

Cullen Murphy pitched in, simply out of the goodness of his generous heart, and it is not an overstatement to say that I couldn't have finished this book without him. Karen Avrich not only shared my interest in the topic, but she dug up a huge chunk of the articles, studies, and books that became its backbone. Catherine Collins worked with me on numerous long afternoons—punctuated by "catered" shrimp salads—as I tried to impose a structure on my unruly ideas.

From the White House to the carpool line to the many places in between, my friends read, listened, and encouraged me in ways that meant more to me than they'll ever know: Allison Abner, Susan Brophy, Nina Burleigh, Wendy Button, Lisa Caputo, Tim and Colleen Crescenti (just because), Anne Dickerson, Beth Dozoretz, Tammy Haddad, Capricia Marshall, Jaye Rogovin,

Vicki Rollins, Hilary Rosen, Cathy Saypol, Ricki Seidman, Claire Shipman, Beegie Truesdale, Christine Varney, Lynne Wasserman, and Sharon Zalusky.

My agent, Robert Barnett, helped me get this project off the ground. And the team at HarperCollins was fantastic. My editor, Claire Wachtel, believed in this book—and in me—from Day One; she happily provided a pep talk on the many occasions when I needed one. Thanks, too, to Campbell Wharton, who worked tirelessly to promote the book, and to Julia Novitch, Kathy Schneider, and Tina Andreadis.

And finally, my heartfelt thanks to the amazing women (and a few men) who sat with me and answered my endless questions. I was honored that they shared their time with me and inspired by the grace, wisdom, and good humor with which they shared their stories: Senator Dianne Feinstein, the Honorable Geraldine Ferraro, Jane Friedman, Robin Gerber, Jane Goodall, Dr. Bernadine Healy, Secretary Alexis Herman, Ambassador Swanee Hunt, Senator Kay Bailey Hutchison, Sherry Lansing, Debra Lee, Laura Liswood, Wangari Maathai, Judith McHale, Pat Mitchell, Tom Peters, the late Anita Roddick, Governor Kathleen Sebelius, Don Steinberg, Dr. Shirley Tilghman, Ken Wallach, and Marie Wilson.

Notes

Introduction

2 A comprehensive review of encyclopedia entries: Karin Klenke, *Women and Leadership: A Contextual Perspective* (New York: Springer Publishing, 1996).

7 "We relate on a personal level, because": quoted in Mary Leonard, "Transfer of Power: The Female Contingent," *Boston Globe*, January 19, 2001.

7 the Irish women came away inspired: Susan Page, "Across Party Lines, Senate's Women Forge Unique Bonds," *USA Today*, July 24, 2000.

7 "women tend to be better at working across the aisles": quoted in Lizette Alvarez, "Feminine Mystique Grows in Senate," *New York Times*, December 7, 2000.

9 Still, women make up only 16 percent: Center for American Women and Politics, Rutgers University, *Women in Elected Office 2007 Fact Sheet*. Viewed online at http://www.cawp.rutgers.edu/Facts/Officeholders/elective.pdf.

9 women account for only 16 percent: Catalyst, *2006 Catalyst Census of Women Corporate Officers and Top Earners of the Fortune 500*. Viewed

online at http://www.catalyst.org/knowledge/titles/title.php?page=cen_COTE_06.

9 Women make up half of law school graduates: Deborah Rhode, *The Unfinished Agenda: Women and the Legal Profession*, The American Bar Association Commission on Women in the Profession, 2001. Viewed online at http://www.abanet.org/ftp/pub/women/unfinishedagenda.pdf.

9 Women make up nearly half of medical school graduates: Bonnie Darves, "Women in Medicine Force Change in the Workforce Dynamic," *New England Journal of Medicine*, April 2005. Viewed online at http://www.nejmjobs.org/career-resources/women-in-medicine.aspx.

10 They are 20 percent of university presidents: Caryn McTighe Musil, "Harvard Isn't Enough," *Ms.*, April 2007.

CHAPTER 1: BETWEEN A POLITICAL ROCK AND A PROMISE

15 And after checking old clips: Richard L. Berke, "Clinton Selects a Mostly Youthful Group of White House Aides, *New York Times*, January 15, 1993.

20 Clinton found himself playing defense: Catherine S. Manegold, "Clinton Ire on Appointments Startles Women," *New York Times*, December 23, 1992.

24 In fact, women devalue whole sectors: Catalyst, *The Double Bind Dilemma for Women in Leadership: Damned If You Do, Doomed If You Don't*, 2007.

24 In 2002, 40 percent of medical residents: Bonnie Darves, "Women in Medicine Force Change in Workforce Dynamics," *New England Journal of Medicine*, April 2005.

24 their broadcasts haven't shown any real ratings growth: According to *Broadcasting and Cable*, ABC's *World News*, anchored by Charlie Gibson, averaged 8.95 million viewers the week ending October 26, 2007. While that was better than NBC's *Nightly News* anchored by Brian Williams (which averaged 8.51 million viewers) or CBS's *Evening News* anchored by Katie Couric, it marked the broadcast's

biggest audience in eight months (since February 2007). The numbers dropped again in subsequent weeks. *Broadcasting and Cable* information viewed online at http://www.broadcastingcable.com/article/CA6495665.html.

27 women working full-time earned just eighty cents: U.S. Government Accounting Office, *Women's Earnings: Work Patterns Partially Explain Difference between Men's and Women's Earnings*, October 2003. Viewed online at http://www.gao.gov/new.items/d0435.pdf.

27 even when you control for other factors: Hillary Lips, "Women, Education, and Economic Participation," Keynote Address Presented at the Northern Regional Seminar, National Council of Women of New Zealand, Auckland, New Zealand, March 1999.

27 "people think that what men do is more important": quoted in Betsy Morris, "How Corporate America Is Betraying Women," *Fortune*, January 10, 2005.

28 Fifty-five percent . . . said they wanted to be CEOs: Catalyst, *Women "Take Care," Men "Take Charge": Stereotyping of U.S. Business Leaders Exposed*, 2005.

28 An executive at Boeing told *Fortune*: Morris, "How Corporate America Is Betraying Women."

29 men negotiated their initial salary: Linda Babcock and Sara Laschever, *Women Don't Ask: Negotiation and the Gender Divide* (Princeton: Princeton University Press, 2003), viewed online at www.womendontask.com/stats.html.

29 failing to negotiate that first salary: ibid.

29 "Society really teaches young girls": Linda Babcock, interviewed by Bill Hemer for CNN, August 21, 2003.

31 "We hear that it's in its final phases": Dee Dee Myers, White House Briefing, June 25, 2003 (transcript).

33 By virtually every measure, the assault was a success: Eric Schmitt, "U.S. Says Strike Crippled Iraq's Capacity for Terror," *New York Times*, June 28, 1993.

34 White House's success in keeping the missile strike secret: Doug Jehl, "Administration Finds Just Keeping a Secret Can Be a Triumph," *New York Times*, June 28, 1993.

35 "Dee Dee, you've been reported to be concerned": Dee Dee Myers, White House Briefing, June 28, 1993 (transcript).

35 Larry Speakes, President Reagan's spokesman: Larry Speakes, *Speaking Out* (New York: Charles Scribner's Sons, 1988), 150–54.

36 Larry King did a show: Brit Hume, Andrea Mitchell, Rita Braver, and Wolf Blitzer interview by Larry King, *Larry King Live*, CNN.

CHAPTER 2: WHY CAN'T A WOMAN (BE MORE LIKE A MAN)

41 perceive women as better at "caretaking skills": Catalyst, *Women "Take Care," Men "Take Charge."*

43 she and her husband had hired a Peruvian couple: Robert Reinhold, "An Angry Public, Fueled by Radio and TV, Outcry Became Uproar," *New York Times*, January 23, 1992.

44 "Voters focus on a female candidate's performance": *Cracking the Code: Political Intelligence for Women Running for Governor*, The Barbara Lee Family Foundation, 2004, 21.

45 if a woman is too assertive: Alice H. Eagly and Linda L. Carli, "Women and the Labyrinth of Leadership," *Harvard Business Review*, September 2007.

45 "I think [men] won't tolerate some things": Kathleen Sebelius, interview by the author, September 13, 2006.

46 "What I hear from young women to the old": Anonymous interview by the author, May 16, 2006.

47 "It's not easy": YouTube, viewed online at http://fr.youtube.com/watch?v=pl-W3IXRTHU&feature=dir.

48 "Nothing drives me crazier": Judith McHale, interview by the author, July 18, 2006.

48 "Why do you feel that way?": ibid.

49 "The emphasis on Katie's appearance": quoted in Howard Kurtz, "At CNN, Taking On the Cable Guys," *Washington Post*, July 30, 2007.

53 when someone tries to "restate" one of their ideas: Catalyst, *The Double Bind Dilemma*, 28.

54 on the CBS Web site, Couric recounted: Katie Couric, "A Woman

at the Table," viewed online at http://www.cbsnews.com/blogs/2007/01/17/couricandco/entry2366267.shtml.

55 studied 15,000 men—and no women: Malcolm Gladwell, "The Healy Experiment," *Washington Post Magazine*, June 21, 1992.

55 government take women's health issues more seriously: Erik Eckholm, "A Tough Case for Dr. Healy," *New York Times Magazine*, December 1, 1991.

56 "So when I got there, I said, 'Guess what?'": Bernadine Healy, interview by the author, July 13, 2006.

56 Healy launched the Women's Health Initiative: Eckholm, "A Tough Case for Dr. Healy."

56 until recently, *all* the research into the disease: U.S. Department of Health and Human Services, National Institutes of Health, *Subtle and Dangerous: Symptoms of Heart Disease in Women*, 2006.

57 "everything else becomes men are the normative standard": Healy interview.

57 women are the engine driving economic growth worldwide: "The Importance of Sex," *Economist*, April 12, 2006.

58 companies with the highest representation of women: Catalyst, *The Bottom Line: Corporate Performance and Women's Representation on Boards*, October 1, 2007.

58 Women now earn 60 percent: Tamar Lewin, "At Colleges, Women Are Leaving Men in the Dust," *New York Times*, July 9, 2006.

58 Women already make the vast majority: Hillary Chura, "Failing to Connect: Marketing Messages for Women Fall Short," *Advertising Age*, September 23, 2002.

58 women will control some $12 million: Maureen Nevin Duffy, "Tips for Working with Female Clients: Women Approach Investing Differently than Men," *Journal of Accountancy*, April 1, 2004.

59 The Right Hand Rules the World: Tim Schooley, "Does De Beers Ad Campaign Ring True?" *Pittsburgh Business Times*, February 27, 2004.

59 devised a basket of 115 Japanese companies: "A Guide to Womenomics," *Economist*, April 12, 2006.

59 women . . . are also wise investors: "The Importance of Sex," and "Merrill Lynch Investment Managers Survey Finds: When It Comes

to Investing, Gender a Strong Influence on Behavior," *Business Wire*, April 18, 2005.

59 In every presidential election since 1980: Center for American Women in Politics, Rutgers University, *The Gender Gap: Voter Choices in Presidential Elections*, March 2005.

60 women are more likely to introduce and support: Michele Swers, "Understanding the Policy Impact of Electing Women: Evidence from Congress and State Legislatures," PS*: Political Science and Politics*, Vol. 34, No. 2, June 2001.

60 women are "better listeners, more honest and can work across party lines": Deborah Barfield Berry, "Seasoned Women Face Critical Races, *USA Today*, August 17, 2006.

60 Nobel Prize laureate economist Amartya Sen: Amartya Sen, *Development as Freedom* (New York: Anchor Books, 1999), 203.

CHAPTER 3: BIOLOGY, IDEOLOGY, AND DIFFERENCE

64 most women want pretty much the same things: Danielle Crittenden, *What Our Mother's Didn't Tell Us* (New York: Simon & Schuster, 2000), 23.

64 "I'm not a women's lib person or anything": quoted in Sean Gregory, "It's Ladies Fight," *Time*, July 30, 2007.

66 "Here's the president of Harvard": quoted in K. C. Cole, "Sally Ride," *Smithsonian*, November 2005.

67 "Larry put his foot in it": Shirley Tilghman, interview by the author, June 20, 2006.

68 there is no gender difference: Steven Pinker and Elizabeth Spelke, *The Harvard Debate*, Mind/Brain/Behavior Institute, Harvard University, April 22, 2005.

68 significant difference in "variation" : ibid.

70 99 percent of our genetic material: U.S. Department of Health and Human Services, National Institutes of Health, *New Genome Comparison Finds Chimps, Humans Very Similar at DNA Level*, August 31, 2005, viewed online at http://www.genome.gov/15515096 and William Saletan, "Don't Worry Your Pretty Little Head: The Pseudo-

Feminist Show Trial of Larry Summers," *Slate*, January 21, 2005, viewed online at http://slate.com/id/2112570/.

70 "the brain is a sex organ": quoted in Amanda Ripley, "Who Says a Woman Can't Be Einstein," *Time*, March 7, 2005.

70 male and female brains are indistinguishable: Louann Brizendine, *The Female Brain* (New York: Morgan Road Books, 2007), 14.

71 women have on average 11 percent more neurons: ibid., 5.

71 much less debilitating in a woman: Simon Baron-Cohen, *The Essential Difference* (London: Penguin Books, 2004), 106, and Dorian Sagan, "Gender Specifics: Why Women Aren't Men," *New York Times* on the Web, June 21, 1998, viewed online at http://www.nytimes.com/specials/women/nyt98/21saga.html.

71 "peripheral visionaries": quoted in Sagan, "Gender Specifics."

71 these differences manifest themselves in a variety of ways: Baron-Cohen, *Essential Difference*, 32.

72 female brains are wired to empathize: Ibid., 31–32.

73 ability to recognize emotion: Ibid., 59 and Appendix 2.

73 Stossel's crew made lemonade: John Stossel, "The Difference Between Boys and Girls," ABC News, April 14, 2006, viewed online at http://www.abcnews.go.com/2020/print?id=123726.

74 "I didn't want to make anyone feel bad": ibid.

74 what happens in the brain when someone falls in love?: "How the Brain Reacts to Romance," BBC News, November 12, 2003, viewed online at http://newsvote.bbc.co.uk./mpapps/pagetools/pringt/news.bbc.co.uk/1/hi/health/3261309.stm.

75 men laugh at flatulence jokes: quoted in "She Who Laughs Last," *The Week*, November 25, 2005.

75 intimacy is the fabric of relationships: Deborah Tannen, "Sex, Lies and Conversation: Why Is It So Hard for Men and Women to Talk to Each Other?" *Washington Post*, June 24, 1990.

75 "a trip to the surface of the Moon": quoted in "What are the 78 Differences Between Women and Men?" BBC News, June 19, 2003, viewed online at http://news.bbc.co.uk/2/hi/uk_news/3002946.stm.

76 "at every age, the girls and women faced each other": ibid.

76 men's self-esteem derives more: Brizendine, *Female Brain*, 41.

77 "women were kept out of the legal profession": in "If Women Ruled the World: A Washington Dinner Party," *The Millennium Dinners*, PBS, first aired March 17, 2001.

77 "hysteria": Barbara Ehrenreich and Deirdre English, *For Her Own Good: 150 Years of the Experts' Advice to Women* (New York: Anchor, 2005); Rachel F. Maines, *The Technology of Orgasm: "Hysteria," the Vibrator, and Women's Sexual Satisfaction* (Baltimore: Johns Hopkins University Press, 2001); and Bill Cassleman, *Bill Casselman's Dictionary of Medical Derivations* (Pearl River, NY: Parthenon Publishing Group, 1998).

78 "Danica is pretty aggressive in our cars": quoted in Perspectives, *Newsweek*, July 24, 2006.

79 when infant monkeys are given a choice: Ronald Kotulak, "Gender and the Brain," *Chicago Tribune*, April 30, 2006.

79 neutralize traditional gender roles: Lawrence H. Summers, Remarks at NBER Conference on Diversifying the Science and Engineering Workforce, Cambridge, MA, January 14, 2005, viewed online at http://www.president.harvard.edu/speeches/2005/nber.html.

79 where biological boys have been raised as girls: John Colapinto, *As Nature Made Him: The Boy Who Was Raised a Girl* (New York: HarperPerennial, 2006).

80 "what if the communication center": Brizendine, *Female Brain*, 8.

81 "Women have a biological imperative": Ibid., 163.

CHAPTER 4: IF THE THREE WISE MEN HAD BEEN WOMEN

86 when the women who worked in the lab: "UCLA Researchers Identify Key Biobehavioral Pattern Used by Women to Manage Stress," *Science Daily*, May 22, 2000, and Gale Berkowitz, "UCLA Study on Friendship Among Women," 2002, viewed online at http://www.anapsid.org/cnd/gender/tendfend.html.

88 basket neatly stacked with free tampons: Yilu Zhao, "Beyond Sweetie," *New York Times*, November 7, 2004.

88 Sherry Lansing, the former chief of Paramount: Sherry Lansing, interview by the author, August 22, 2006.

90 "if 50 percent of your viewing public": McHale interview.

90 Look at Revlon: Tom Peters, interview by the author, May 11, 2006; Revlon Web site viewed online at http://www.corporate-ir.net/ireye/ir_site.zhtml?ticker=REV&script=2200; and Procter & Gamble Web site viewed online at http://phx.corporate-ir.net/phoenix.zhtml?c=195341&p=irol-govmanage.

91 a speech he gave at a gem . . . conference: Peters interview.

91 Debra Lee, the CEO of Black Entertainment Networks: Debra Lee, interview by the author, May 25, 2006.

93 both Republican and Democratic women: *Cracking the Code*, 2.

93 Kay Bailey Hutchison was the first: Kay Bailey Hutchison, interview by the author, July 12, 2006.

93 Hutchison believes: ibid.

94 she's a "huge believer": Sebelius interview.

95 "My mother wrote that law!": Lindy Boggs, *Washington through a Purple Veil: Memoirs of a Southern Woman* (New York: Harcourt Brace & Company, 1994), 312–14.

96 "I listened to the women from the countryside": Wangari Maathai, Speech to Accenture, Washington DC, May 24, 2006.

97 when women control the family funds: Isobel Coleman, "The Payoff from Women's Rights," *Foreign Affairs*, May/June 2004.

97 in India, researchers found: Raghabendra Chattopadhyay and Ester Dufflo, *Women as Policy Makers: Evidence from an India-wide Randomized Policy Experiment*, Massachusetts Institute of Technology, October 2001.

98 motherhood literally alters a woman's brain: Brizendine, *Female Brain*, 95.

99 the constant physical and emotional demands: Katherine Ellison, *The Mommy Brain: How Motherhood Makes Us Smarter* (New York: Basic Books, 2005).

101 "The expression 'the buck stops here'": Daniel Stern, *The Birth of a Mother* (New York: Basic Books, 1998), 15–17.

102 when she became president of Princeton: Tilghman interview.

103 "Give a woman an inch, and she'll park a car in it": Geraldine Ferraro, Remarks to the Madeleine Albright Grants Luncheon, Na-

tional Democratic Institute, Washington DC, May 1, 2006.

103 "Women are accustomed, I guess, to cleaning up": Dianne Feinstein, interview by the author, July 28, 2007.

CHAPTER 5: THE NATURE OF VIOLENCE

107 women hold thirty-nine out of eighty seats: Inter-Parliamentary Union, *Women in National Parliaments*, October 31, 2007, viewed online at http://www.ipu.org/wmn-e/classif.htm.

108 "Before the genocide": quoted in Marc Lacy, "Women's Voices Rise and Rwanda Reinvents Itself," *New York Times*, February 26, 2005.

108 "We were the wives left as widows": quoted in Emily Flynn Vencat, "Keepers of the Peace," *Newsweek*, November 14, 2005.

108 "It was like asking Jews": Swanee Hunt, interview by the author, July 20, 2006.

108 "Widows of the genocide": quoted in "Women Taking a Lead: Progress Toward Empowerment and Gender Equity in Rwanda," *Critical Half: Bi-Annual Journal of Women for Women International*, September 2004, 15.

109 women became symbols of moderation: Elizabeth Powley, *Strengthening Governance: The Role of Women in Rwanda's Transition*, United Nations Office of the Special Advisor on Gender Issues, January 26, 2004, 4.

109 "women are better at reconciliation": quoted in Lacy, "Women's Voices Rise."

109 "Traditionally, a woman is not a breadwinner": quoted in Mike Crawly, "Rwandan Social Structure Evolves," *Christian Science Monitor*, June 21, 2000.

109 30 percent of the seats . . . were set aside for women: Lacy, "Women's Voices Rise."

110 "Some men even complained": ibid.

110 "women rolled up their sleeves": quoted in *The Vital Role of Women in Peace Building*, Hunt Alternatives Fund, viewed online at http://www.huntalternatives.org/pages/460_the_vital_role_of_women_in_peace_building.cfm.

110 "women have served as peace educators": quoted ibid.

110 women are . . . better at creating and keeping the peace: Vencat, "Keepers of the Peace."

111 "After the war in Bosnia": Hunt interview.

111 "if mothers ruled the world": quoted in James Hibberd, "Sally Field Speaks Out at the Emmys," *TV Week*, September 16, 2007, viewed online at http://www.tvweek.com/blogs/james-hibberd/2007/09/sally_field_at_the_emmys.php.

112 "essential" that women participate in every phase: Klara Banaszak, Camille Pampell Conaway, Anne Marie Goetz, Aina Iiyambo, and Maha Muna (editors), *Securing the Peace: Guiding the International Community towards Women's Effective Participation throughout Peace Processes*, United Nations Development Fund for Women, October 2005.

112 "Women are more practical": Mostar women, interview by the author, May 1, 2006.

113 "Men are stubborn": quoted in Vencat, "Keepers of the Peace."

113 During South Africa's post-apartheid transition: Pumla Gobodo-Madikizela, *Women's Contributions to South Africa's Truth and Reconciliation Commission*, Women Waging Peace Policy Commission, February 2005.

114 "I convinced Thomas Friedman": Pat Mitchell, interview by the author, May 16, 2006.

115 "the voice of women is clearer": Shulamit Aloni, "Give Peace a Chance: Women Speak Out," bitterlemons-international.org, September 4, 2003, viewed online at http://www.bitterlemons-international.org/previous.php?opt=1&id=9#37.

116 "During my meetings with Palestinian women": quoted in Anat Cohen, "Israeli and Palestinian Women Talk About Peace," Women's eNews, March 11, 2003, viewed online at http://www.womensenews.org/article.cfm?aid=1251.

116 "For men, negotiation": quoted ibid.

116 "women usually delve into the little details": quoted ibid.

117 recent efforts to empower women: Sen, *Development as Freedom*, 178–79.

118 educating girls probably produced better returns: Coleman, "The Payoff."

118 "increases in household income": ibid.

119 large gap in the literacy rates: Steven Fish, "Islam and Authoritarianism," *World Politics*, October 2002.

119 "One time I found a woman": quoted in Richard Wrangham and Dale Peterson, *Demonic Males* (Boston: Houghton Mifflin, 1996), 4.

120 men are the primary perpetrators: ibid., 113.

121 chimpanzees engage in certain behaviors: Esther Addley, "Life: The Ascent of One Woman," *Guardian*, April 3, 2003.

121 the group split into two separate factions: Wrangham and Peterson, *Demonic Males*, 12–16.

122 "It was only one female who really participated": Jane Goodall, interview by the author, September 15, 2006.

122 "It was a very primitive war": ibid.

122 a male chimp in his prime: Wrangham and Peterson, *Demonic Males*, 191.

122 Of the more than 4,000 mammals: ibid., 23–24.

123 "Male chimps are very aggressive": Goodall interview.

123 "phenomena like aggression": Francis Fukuyama, "Women and the Evolution of World Politics," *Foreign Affairs*, September/October 1998.

125 Inter-Parliamentary Union's annual ranking: Inter-Parliamentary Union, *Women in National Parliaments*.

128 "chimpanzees show love": quoted in Addley, "Life."

CHAPTER 6: GETTING TO WIN-WIN

129 an ominous feeling: Alexis Herman, interview by the author, August 1, 2006.

130 One hundred eighty-three thousand workers: Charles Krause, "Package Deal," *The NewsHour*, PBS, August 19, 1997.

132 "absolutely crucial" to the settlement: quoted in Elizabeth Shogren, "The Strike Settlement: Herman Survives Trial by Fire," *Los Angeles Times*, August 20, 1997.

133 rated her performance "at least a 9, if not a 9.5": quoted in Kevin

Merida, "For Alexis Herman, a Proving Ground," *Washington Post*, August 20, 1997.

133 Is there such a thing as a "female style"?: Alice Eagly, Mary Johannesen-Schmidt, Marloes van Engen, Transformational, *Transactional and Laissez-Faire Leadership Styles: A Meta-Analysis Comparing Men and Women*, Institute for Policy Research at Northwestern, 2004.

134 "By valuing a diversity of leadership styles": Judith Rosener, "Ways Women Lead," *Harvard Business Review*, November/December 1990.

134 "forced to pioneer . . . strategies": Sally Helgesen, "Women and the New Economy," *Leader to Leader*, 1997, viewed online at http://www3.interscience.wiley.com/cgi-bin/abstract/114100179/ABSTRACT?CRETRY=1&SRETRY=0.

135 "When you put together a thirty-person project team": Peters interview.

136 "This is why you want to hire women": Pat Mitchell interview.

136 When Kathleen Sebelius was insurance commissioner: Sebelius interview.

137 "All the women leaders I've met": quoted in "When Women Lead," *Newsweek*, October 24, 2005.

138 "I do think there's such a thing as intuition": Feinstein interview.

138 "I think we make them faster": Jane Friedman, interview by the author, May 15, 2006.

138 "I'm not for a woman in any government job": quoted in Adam Clymer, "Book Says Nixon Considered a Woman for the Supreme Court," *New York Times*, September 27, 2001.

139 Gut feelings are not just: Brizendine, *Female Brain*, 120.

139 "women are better than men at decoding": quoted in David G. Myers, "The Powers and Perils of Intuition," *Psychology Today*, November-December 2002.

139 "It's only when men actually see tears": Brizendine, *Female Brain*, 119.

141 To forge connection: ibid., 28–29.

141 "The advice from the men": Peters interview.

141 "I do think women approach things": Sebelius interview.

141 "Always leave a little something": quoted in "12 Leaders on Life Lessons," *Newsweek*, October 24, 2005.

142 "'No, it's got to be my baby'": Sebelius interview.

142 "Women are more honest": Lee interview.

143 They call themselves "the Sirens": Karen Breslau, "A New Team in Town," *Newsweek*, October 24, 2005.

144 Tactics that aren't "badge heavy": ibid.

144 "I do think mostly women are more collaborative": Kay Bailey Hutchison, interview by the author, July 12, 2006.

145 "excited by an interdisciplinary approach": quoted in Victoria Griffith, "Overhaul of a Grand Institution," *Financial Times*, March 7, 2005.

145 "I seek out people": quoted in Jonathan Darman, "A Much Fuller Understanding," *Newsweek*, September 17, 2007.

146 "They don't like power": Anita Roddick, interview by the author, September 18, 2006.

147 women leaders are increasingly speaking of mothering: Erkut, *Inside Women's Power.*

147 "It's a sign of their comfort with motherhood": quoted in Mary Meier, "Leaders Say Managing Kids Prepared Them to Be Boss," Women's eNews, October 16, 2001, viewed online at http://www.womensenews.org/article.cfm/dyn/aid/687.

147 "better leader in my job because I'm a parent": quoted in "12 Leaders on Life Lessons."

147 "It's about consistent discipline": Breslau, "A New Team in Town."

148 "I always find it very difficult to say": Wangari Maathai, interview by the author, May 24, 2006.

148 Judith McHale also expressed doubts: McHale interview.

148 "combine a strong focus on results": Erkut, *Inside Women's Power.*

149 "less effective and less fair than males": Leanne E. Atwater, James A. Carey, and David A. Waldman, "Gender and Discipline in the Workplace: Wait Until Your Father Gets Home," *Journal of Management*, September 1, 2001.

149 "when it comes to do the obituary on me": Roddick interview.

CHAPTER 7: PLUGGING THE LEAKY PIPELINE

155 "Jane Swift had twins": quoted in Rita Braver, "The Quest to

Become Ms. President," *CBS News Sunday Morning*, February 5, 2006.

156 "Waiting for the connection between gender": Susan Estrich, *Sex & Power* (New York: Riverhead Books, 1999).

156 women cite lack of opportunity: Peninah Thomson and Jacey Graham, *A Woman's Place Is in the Boardroom* (Houndsmills, England: Palgrave Macmillan, 2005).

156 These "micro-inequities" are like pebbles: Alison Maitland, "The Hidden Obstacles to Women's Final Ascent," *Financial Times*, September 12, 2005.

156 the accounting and consulting giant Deloitte & Touche: Douglas M. McCracken, "Winning the Talent War for Women: Sometimes It Takes a Revolution," *Harvard Business Review*, November/December 2000.

157 men were considered better musicians than women: Malcolm Gladwell, *Blink* (New York, Little, Brown and Company, 2005), 250.

158 Two different résumés were used: Pinker and Spelke, *The Harvard Debate*.

158 Nancy Hopkins . . . famously showed: Committee on Women Faculty, *A Study on the Status of Women Faculty in Science at MIT*, Massachusetts Institute of Technology, 1999.

159 "I always believed that contemporary gender discrimination": Charles M. Vest, Introductory Comments, *A Study on the Status of Women Faculty in Science at MIT*.

159 "If you're in pediatrics": Healy interview.

160 "Men are willing to talk about these things": quoted in Jody and Matt Miller, "Get a Life," *Fortune*, November 28, 2005.

162 desire to work with "people versus things" : Pinker and Spelke, *The Harvard Debate*.

163 viewed "pure math and physics careers": Laura Vanderkam, "What Math Gender Gap?" *USA Today*, April 12, 2005.

163 girls who scored extremely high on the math portion: ibid.

164 As opportunities for girls to play skyrocketed: Terri Lakowski, "Title IX Myth-Fact," viewed online at http://www.womenssportsfoundation.org/cgi-bin/iowa/issues/rights/article.html?record=1209.

164 why there are more women in her field: Tilghman interview.

165 "I think there are a whole series of things": ibid.

166 was a champion of flexibility: McHale interview.

167 "You can have flexible hours": Hutchison interview, July 12, 2006.

167 family-friendly policies increased: Alice Eagly and Linda L. Carli, "Women and the Labyrinth of Leadership," *Harvard Business Review*, September 2007.

168 The survey of mid-career women: Sylvia Ann Hewlett, "Off Ramps and On Ramps: Keeping Talented Women on the Road to Success," *Harvard Business Review*, September/October 2005.

168 "all you needed to do was fill the pipeline": quoted in Daniel McGinn, "Getting Back on Track," *Newsweek*, September 25, 2006.

168 The first time I opened *Newsweek*: Advertisement following page 61, *Newsweek*, July 17, 2006.

169 "whoever figures this out first wins": quoted in McGinn, "Getting Back on Track."

169 Domino's pizza found: *Wall Street Journal*, February 2005.

169 "My kids were little": Sebelius interview.

170 2.5 percent of babies: according to various sources, including the Centers for Disease Control, about 4.1 million babies were born in the United States in 2005. Roughly 110,000—or 2.6 percent—of those were born to women forty or older.

170 "recognizing that fifty is probably the new thirty": Sebelius interview.

171 "I could not have done my job and raised children": Lansing interview.

172 "left my job as a correspondent for *60 Minutes*": quoted in "Lessons We Have Learned," *Newsweek*, September 25, 2006.

173 Hillary Clinton said she'd been touched: Anne E. Kornblut, "Encouraged by Women's Response, Clinton Stresses Female Side," *Washington Post*, October 7, 2007.

CHAPTER 8: CLOSING THE CONFIDENCE GAP

177 He recounted that I had urged him to talk to the press: Bill Clinton, *My Life* (New York: Alfred A. Knopf, 2004), 499.

178 girls are more comfortable making decisions together: Brizendine, *Female Brain*, 22.

179 "The women are very bright": quoted in Daniel McGinn, "Vote of Confidence," *Newsweek*, October 24, 2005.

179 "Middle school is the moment of bifurcation": quoted in *Research & Action Report*, Wellesley Centers for Women, Spring/Summer 2005, 10–11.

180 "stereotype threat": Claude Steele and Joshua Aronson, "Stereotype Threat and the Intellectual Test Performance of African Americans," *Journal of Personality and Social Psychology*, November 1995.

180 Can you override stereotypes: Mathew McGlone and Joshua Aronson, "Social Identity Salience and Stereotype Threat in Standardized Test Performance," Conference Paper, International Communications Association Conference, New Orleans, May 2004.

180 "We were activating their snob schema": quoted in Richard Morin, "Women Aren't Good in Math . . . Or Are They?" *Washington Post*, August 31, 2006.

181 "Women negotiate very effectively": quoted in Dawn Klingensmith, "Mapping Your Route to the Top Job," *Chicago Tribune*, June 8, 2005.

182 men think of negotiating like a ballgame: Babcock and Laschever, *Women Don't Ask*.

182 "women need to find their own negotiating voices": Babcock interview by Bill Hemer.

183 "Women tend to run because": quoted in "Women's March into Office Slows," *Wall Street Journal*, August 16, 2007.

183 women are even less likely to run: Jennifer Lawless and Richard L. Fox, *Why Don't Women Run for Office*, Taubman Center for Public Policy, Brown University, January 2004.

183 "I knew how we could win": Sebelius interview.

184 "There aren't enough women who say": Hunt interview.

185 "I'd go straight to Mrs. Truman and apologize": quoted in Richard Sevaro, "Margaret Chase Smith Is Dead at 97; Maine Republican Made History Twice," *New York Times*, May 30, 1995.

186 "Don't just think of me as a lawyer": McHale interview.

187 "improve the species": The Darwin Awards Web site, viewed online at http://www.darwinawards.com/darwin/.

187 major financial houses did a survey: Olivia Millan and Karina Piskaldo, "Men, Women and Money," *Psychology Today*, January 1999.

188 "women engage in . . . 'protective hesitation'": Herman interview.

190 freshmen . . . awarded . . . Westinghouse Science Prize: Tilghman interview.

190 fathers get more credit than mothers: Deborah Rhode, *Speaking of Sex* (Cambridge, MA: Harvard University Press, 1997).

191 "these boys who get C's in math": quoted in Cole, "Sally Ride."

191 "When it comes to professional modesty": quoted in Patricia Sellers, "Power: Do Women Really Want It?" *Fortune*, October 13, 2003.

192 "weren't a lot of female breast surgeons": quoted in "12 Leaders on Life's Lessons."

192 "Don't assume that others are aware": ibid.

192 "Bobby is going to say he's responsible": quoted in George Vecsey, "People's Champ Earns Public Tribute," *New York Times*, August 29, 2006.

193 "I did my reports": Lee interview.

193 "One of Hollywood's first successful female producers . . . that's what I can't figure out": Anonymous interview by the author, March 12, 2007.

194 "When I made a mistake in the lab": Tilghman interview.

CHAPTER 9: SEEING IS BELIEVING

197 *If the Walls Could Talk*: Jane O'Connor and Gary Hovland, *If the Walls Could Talk: Family Life at the White House* (New York: Simon & Schuster, 2004).

205 When I talked to Geraldine Ferraro: Geraldine Ferraro, interview by the author, May 15, 2006.

206 "The women licked the envelopes": Sebelius interview.

207 "There will never be a woman head of studio": Lansing interview.

207 Hollywood's "immigrant, outsider ethos": quoted in Nancy Hass, "Hollywood's Old Girls' Network," *New York Times*, April 24, 2005.

208 "The notion that you can't see anybody": Tilghman interview.
212 Louis Leakey . . . believed that women: Addley, "Life."
212 "After every single lecture": Goodall interview.
213 "It was in college, through my books": Herman interview.
213 "I found a biography of Maria Mitchell": quoted in "When Women Lead," *Newsweek*, October 24, 2005.
213 "I'll never forget when I was in the sixth grade": Hutchison interview, July 12, 2006.
215 "No longer should the question be": Doug Burns, "Clinton's Debate Dominance Flips Script: Why Hasn't a Woman Been President?" *Iowa Independent*, July 23, 2007.
216 In 1967, married women in Texas: Elizabeth York Enstam, *Handbook of Texas Online*, Texas State Historical Association, viewed online at http://www.tsha.utexas.edu/handbook/online/articles/WW/jsw2.html.
217 "As I go by shaking hands and meeting people": quoted in Kornblut, "Encouraged by Women's Response."

CHAPTER 10: REACHING CRITICAL MASS

219 I opened my last briefing: Dee Dee Myers, White House Briefing, December 22, 1994 (transcript).
222 when word of the meeting leaked: Jeffrey Birnbaum, *Madhouse: The Private Turmoil of Working for the President* (New York: Times Books, 1996), 184.
226 Frances Kaiser, the sheriff of Kerr County, Texas: in "If Women Ruled the World: A Washington Dinner Party."
227 "In politics, if you want anything said": Margaret Thatcher, Brainyquote.com, viewed online at http://www.brainyquote.com/quotes/quotes/m/margaretth153838.html.
228 "Her voice and her manner reminded many": R. W. Apple, "Conservatives Win British Vote," *New York Times*, May 4, 1979.
229 "Margaret Thatcher damaged women's place": quoted in Ollie Stone-Lee, "Thatcher's Role for Women," BBC News, November 11, 2005, viewed online at http://news.bbc.co.uk/1/hi/uk_politics/4435414.stm.

229 When she became prime minister in 1979: ibid.

229 Thatcher "created a constituency for women leaders": in "If Women Ruled the World: A Washington Dinner Party."

229 Xerox began planning for succession: Betsy Morris, "Dynamic Duo," *Fortune*, October 15, 2007.

233 "there is strength in numbers": Herman interview.

233 A recent study of corporate boards: Vicki W. Kramer, Alison M. Konrad, and Sumru Erkut, *Critical Mass and Corporate Boards: Why Three or More Women Enhance Governance*, Wellesley Centers for Women, 2006.

234 "In order for [men] to embrace you": Wangari Maathai, Speech to Accenture.

234 Studies show that women begin: Sandra Grey, *Women and Parliamentary Politics*, Australian National University, 2001.

235 Worldwide, only fifteen countries: Inter-Parliamentary Union, *Women in National Parliaments.*

235 Ritt Bjerregaard, the powerful mayor of Copenhagen: Ulla Plon, "Building a Better Family," *Time*, October 12, 2007.

235 Every country that has achieved critical mass: Jane S. Jaquette, "Women in Power: From Tokenism to Critical Mass," *Foreign Policy*, September 22, 1997, and *Getting the Balance Right in National Parliaments*, Women's Environment & Development Organization, 2003, viewed online at http://www.wedo.org/files/5050womnpar.pdf.

236 All of the Nordic countries: Matti Huuhtanen, "Finland Accustomed to Women in Charge; So Are Other Nordic Nations," *Miami Herald*, April 20, 2003.

236 "I certainly prefer": quoted ibid.

236 United Nations evaluates countries: *United Nations Development Program Annual Report*, 2007.

237 "Mary Kay cosmetics": quoted in "Is America Ready for a Woman? Go to Hell," *Huffington Post*, February 2, 2007, viewed online at http://www.huffingtonpost.com/elayne-boosler/is-america-ready-for-a-w_b_40300.html.

237 More women have led governments in South Asia: Vasanthi Ramachandran, "Women Leaders Who Rose to Their Jobs," *New*

Straights Times, November 10, 2003; Neil Western, "Asia's Women Leaders," *Manila Bulletin*, March 7, 2005; and Barbara Crossette, "Enthralled by Asia's Ruling Women? Look Again," *New York Times*, November 10, 1996.

238 Fernandez de Kirchner crushed thirteen opponents: Monte Reel, "Argentina's First Lady Wins Presidency by Wide Margin," *Washington Post*, October 29, 2007.

239 the twenty-first American woman to run: Ramachandran, "Women Leaders Who Rose to Their Jobs."

239 "the trend is genuine and sustained progress": Patricia Sellers, "The Power," *Fortune*, October 15, 2007.

240 "when women become present in greater numbers": in "If Women Ruled the World: A Washington Dinner Party."

Bibliography

Baron-Cohen, Simon. *The Essential Difference.* London: Penguin Books, 2004.

Bennis, Warren. *On Becoming a Leader.* New York: Basic Books, 1989.

Birnbaum, Jeffrey H. *Madhouse: The Private Turmoil of Working for the President.* New York: Times Books, 1986.

Boggs, Lindy. *Washington Through a Purple Veil: Memoirs of a Southern Woman.* New York: Harcourt Brace & Company, 1994.

Brizendine, Louann. *The Female Brain.* New York: Morgan Road Books, 2007.

Buchanan, Bay. *The Extreme Makeover of Hillary (Rodham) Clinton.* Washington, D.C.: Regnery Publishing Inc., 2007.

Collins, Gail. *America's Women: 400 Years of Dolls, Drudges, Helpmates, and Heroines.* New York: William Morrow, 2003.

Coughlin, Linda and Ellen Wingard and Keith Hollihan. *Enlightened Power: How Women Are Transforming the Practice of Leadership.* San Francisco: Jossey-Bass, 2005.

Crittenden, Ann. *If You've Raised Kids, You Can Manage Anything.* New York: Gotham Books, 2004.

———. *The Price of Motherhood.* New York: Owl Books, 2001.

Crittenden, Danielle. *What Our Mothers Didn't Tell Us.* New York: Simon & Schuster Paperbacks, 1999.

Davey, Moira. *Mother Reader: Essential Writings on Motherhood.* New York: Seven Stories Press, 2001.

Edwards, Elizabeth. *Saving Graces.* New York: Broadway Books, 2006.

Ellison, Katherine. *The Mommy Brain: How Motherhood Makes Us Smarter.* New York: Basic Books, 2005.

Ephron, Nora. *I Feel Bad About My Neck.* New York: Alfred A. Knopf, 2006.

Estrich, Susan. *The Case for Hillary Clinton.* New York: Reagan Books, 2005.

———. *Sex & Power.* New York: Riverhead Books, 2000.

Fiorina, Carly. *Tough Choices: A Memoir.* New York: Portfolio, 2006.

Fisher, Helen. *The First Sex: The Natural Talents of Women and How They Are Changing the World.* New York: Random House, 1999.

Friedan, Betty. *The Feminine Mystique.* New York: W. W. Norton & Company, 2001.

Gerber, Robin. *Leadership the Eleanor Roosevelt Way: Timeless Strategies from the First Lady of Courage.* New York: Prentice Hall Press, 2002.

Gilligan, Carol and Janie Victoria Ward and Jill McLean Taylor, *Mapping the Moral Domain.* Cambridge: Harvard University Press, 1988.

Gilligan, Carol, *In a Different Voice.* Cambridge: Harvard University Press, 1982.

Gladwell, Malcolm. *Blink: The Power of Thinking without Thinking.* New York: Little, Brown & Company, 2005.

Gray, John. *Men Are from Mars, Women Are from Venus.* New York: Quill, 1992.

Hankin, Sheenah. *Complete Confidence.* New York: Regan Books, 2004.

Harragan, Betty Lehan. *Games Mother Never Taught You.* New York: Warner Books, 1977.

Helgesen, Sally. *The Female Advantage: Women's Ways of Leadership.* New York: Doubleday Currency, 1995.

Henneberger, Melinda. *If They Only Listened to Us.* New York: Simon & Schuster, 2007.

Hirschman, Linda R. *Get to Work: A Manifesto for Women of the World.* New York: Viking, 2006.

Huffington, Arianna. *On Becoming Fearless.* New York: Little, Brown & Company, 2006.

Hunt, Swanee. *Half Life of a Zealot.* Durham, North Carolina: Duke University Press, 2006.

———. *This Was Not Our War: Bosnian Women Reclaiming the Peace.* Durham, North Carolina: Duke University Press, 2004.

Ingraham, Laura. *The Hillary Trap: Looking for Power in All the Wrong Places.* New York: Hyperion, 2000.

Lake, Celinda and Kellyanne Conway. *What Women Really Want: How American Women Are Quietly Erasing Political, Racial, Class and Religious Lines to Change the Way We Live.* New York: Free Press, 2005.

Lawless, Jennifer L. *It Takes a Candidate.* New York: Cambridge University Press, 2005.

McKenna, Elizabeth Perle. *When Work Doesn't Work Anymore: Women, Work and Identity.* New York: Delacorte Press, 1997.

Milkulski, Barbara et al. *Nine and Counting: The Women of the Senate.* New York: William Morrow, 2000.

O'Beirne, Kate. *Women Who Make the World Worse: How Their Radical Feminist Assault Is Ruining Our Schools, Families, Military and Sports.* New York: Sentinel, 2006.

Olson, Barbara. *Hell to Pay: The Unfolding Story of Hillary Rodham Clinton.* Washington, D.C.: Regnery Publishing Inc., 1999.

Pinker, Steven. *The Blank Slate: The Modern Denial of Human Nature.* New York: Penguin Books, 2004.

Popcorn, Faith. *Evolution: Eight Truths of Marketing to Women.* New York: Hyperion, 1993.

Quinlan, Mary Lou. *Just Ask a Woman: Cracking the Code of What Women Want and How They Buy.* Hoboken, New Jersey: John Wiley & Sons, Inc., 2003.

Rhode, Deborah L. *The Difference Difference Makes: Women and Leadership.* Stanford, California: Stanford Law and Politics, 2003.

Roberts, Cokie. *Founding Mothers: The Women Who Raised Our Nation.* New York: William Morrow, 2004.

———. *We Are Our Mothers' Daughters*. New York: William Morrow, 1998.

Rosener, Judy B. *America's Competitive Secret: Women Managers*. New York: Oxford University Press, 1995.

Sanders, Marlene and Marcia Rock. *Waiting for Prime Time: The Women of Television News*. New York: Harper & Row, 1993.

Shriver, Maria. *Ten Things I Wish I'd Known Before I Went Out into the Real World*. New York: Warner Books, 2000.

Sommers, Christina Hoff. *Who Stole Feminism? How Women Have Betrayed Women*. New York: Simon & Schuster, 1994.

Stassel, Kimberly A. and Celeste Colgan and John C. Goodman. *Leaving Women Behind: Modern Families, Outdated Laws*. Lanham, Maryland: Roman & Littlefield Publishers, Inc., 2006.

Stephanopoulos, George. *All Too Human: A Political Education*. New York: Little, Brown & Company, 1999.

Tannen, Deborah. *That's Not What I Meant! How Conversational Style Makes or Breaks Relationships*. New York: Ballantine Books, 1986.

———. *You Don't Understand! Women and Men in Conversation*. New York: Ballantine Books, 1999.

Tiger, Lionel. *The Decline of Males: The First Look at an Unexpected New World for Men and Women*. New York: St. Martin's Griffin, 1999.

Wilson, Marie C. *Closing the Leadership Gap: Why Women Can and Must Help Run the World*. New York: Viking, 2004.

Wollander, Robin. *Naked in the Boardroom: A CEO Bares Her Secrets So You Can Transform Your Career*. New York: Fireside Books, 2005.

Wrangham, Richard, and Dale Peterson. *Demonic Males: Apes and the Origins of Human Violence*. Boston: Houghton Mifflin Company, 1996.

Index